Fractional Factors and Fractional Deleted Graphs

分数因子和分数消去图

周思中（Sizhong Zhou） 著

图书在版编目(CIP)数据

分数因子和分数消去图 = Fractional factors and fractional deleted graphs:英文/周思中著. —武汉:武汉大学出版社,2014.10
ISBN 978-7-307-14589-4

Ⅰ.分… Ⅱ.周… Ⅲ.图论—研究—英文 Ⅳ.O157.5

中国版本图书馆 CIP 数据核字(2014)第 242304 号

责任编辑:顾素萍　　责任校对:汪欣怡　　版式设计:马　佳

出版发行:武汉大学出版社　(430072　武昌　珞珈山)
　　　　　(电子邮件:cbs22@whu.edu.cn　网址:www.wdp.com.cn)
印刷:湖北睿智印务有限公司
开本:720×1000　1/16　印张:10.25　字数:159 千字　插页:1
版次:2014 年 10 月第 1 版　　2014 年 10 月第 1 次印刷
ISBN 978-7-307-14589-4　　定价:28.00 元

版权所有,不得翻印;凡购我社的图书,如有质量问题,请与当地图书销售部门联系调换。

Contents

Preface 1
Chapter 1 Terminologies and Graphic Parameters 1
   1.1 Basic Terminologies 1
   1.2 Graphic Parameters 2
Chapter 2 Fractional Factors 5
   2.1 Fractional k-Factors 5
   2.2 Fractional k-Factors Including Any Given Edge 48
   2.3 Fractional (g, f)-Factors with Prescribed Properties 57
Chapter 3 Fractional Deleted Graphs 66
   3.1 Fractional k-Deleted Graphs 66
   3.2 Fractional (g, f)-Deleted Graphs 88
Chapter 4 Fractional (k, m)-Deleted Graphs 108
   4.1 A Criterion for Fractional (k, m)-Deleted Graphs 108
   4.2 Degree Conditions for Fractional (k, m)-Deleted Graphs 110
   4.3 Neighborhood and Fractional (k, m)-Deleted Graphs 121
   4.4 Binding Number and Fractional (k, m)-Deleted Graphs 134
   4.5 Toughness and Fractional (k, m)-Deleted Graphs 141
Chapter 5 Fractional (g, f, m)-Deleted Graphs 149
   5.1 Preliminary and Results 149
   5.2 Proof of Main Theorems 151
References 153

Preface

Graph theory is one of the branches of modern mathematics which has shown impressive advances in recent years. Graph theory is widely applied in physics, chemistry, biology, network theory, information sciences, computer science and other fields, and so it has attracted a great deal of attention.

Factor theory of graph is an important branch in graph theory. It has extensive applications in various areas, e.g., combinatorial design, network design, circuit layout, scheduling problems, the file transfer problems and so on.

The fractional factor problem in graphs can be considered as a relaxation of the well-known cardinality matching problem. The fractional factor problem has wide-range applications in areas such as network design, scheduling and combinatorial polyhedra. For instance, in a communication network if we allow several large data packets to be sent to various destinations through several channels, the efficiency of the network will be improved if we allow the large data packets to be partitioned into small parcels. The feasible assignment of data packets can be seen as a fractional flow problem and it becomes a fractional matching problem when the destinations and sources of a network are disjoint (i.e., the underlying graph is bipartite).

In this book, we mainly discuss the fractional factor problem. This book is divided into five chapters.

In Chapter 1, we show basic terminologies, definitions and graphic parameters.

In Chapter 2, we show some sufficient conditions for the existence of fractional factors in graphs. This chapter is divided into three parts. Firstly, we present some sufficient conditions for graphs to have fractional k-factors in terms of neighborhood, binding number, toughness, minimum degree and independence number, etc. It is shown that these results in this part are sharp. Secondly, we investigate the existence of fractional k-factors including any given edge in graphs, and show two results on fractional k-factors including any given edge, and verify that the results are sharp. Thirdly, we study fractional (g, f)-factors with prescribed properties in graphs. We use connectivity and independence number to obtain a sufficient condition for a graph to have a fractional (g, f)-factor, and this result is best possible in some sense. Furthermore, we obtain a result on a fractional (g, f)-factor including any given k edges in a graph.

In Chapter 3, we discuss a generalization of fractional factors, i.e., fractional deleted graphs from different perspectives, such as degree condition, neighborhood condition and binding number. We present some sufficient conditions related to these parameters for the existence of fractional k-deleted graphs and fractional (g, f)-deleted graphs. Furthermore, it is shown that these results are sharp.

In Chapter 4, we study fractional (k, m)-deleted graphs which are the generalizations of fractional k-factors and fractional k-deleted graphs. We first give a criterion for a graph to be a fractional (k, m)-deleted graph. Then we use the criterion to obtain some graphic parameter (such as minimum degree, neighborhood, toughness and binding number, etc.) conditions for graphs to be fractional (k, m)-deleted graphs. Furthermore, it is shown that the results are best possible in some sense.

In Chapter 5, we focus on some sufficient conditions for graphs to be fractional (g, f, m)-deleted graphs. Our results on fractional (g, f, m)-deleted graphs are an extension of the previous results.

The study was supported by the National Natural Science Foundation of China (Grant No. 11371009).

Thanks to Hongxia Liu, Wei Gao, Fan Yang, Yang Xu, Yuan Yuan, Quanru Pan, Jiancheng Wu, who helped to prepare the early drafts of the study and presented numerous helpful suggestions in improving this study.

Preface

I would like to express my gratitude to my family for their encouragement, support and patience when I carried out this study.

Sizhong Zhou

Oct. 6th, 2014

Chapter 1
Terminologies and Graphic Parameters

In this chapter, some basic terminologies, definitions and graphic parameters are given, which will be used throughout this book.

1.1 Basic Terminologies

All graphs considered in this work are finite undirected graphs which have neither loops nor multiple edges. Let G be a graph. We denote by $V(G)$ and $E(G)$ its vertex set and edge set, respectively. A graph H is called a **subgraph** of G if $V(H) \subseteq V(G)$ and $E(H) \subseteq E(G)$. A subgraph H of G is called a **spanning subgraph** of G if $V(H) = V(G)$. For $x \in V(G)$, we use $N_G(x)$ to denote the neighborhood of x in G and $d_G(x)$ to denote the degree of x in G. For any $X \subseteq V(G)$, we define

$$N_G(X) = \bigcup_{x \in X} N_G(x).$$

Note that $N_G(x)$ does not contain x, but it may happen that $N_G(X) \supseteq X$. Let S be a subset of $V(G)$. Denote by $G - S$ the subgraph obtained from G by deleting the vertices in S together with their incident edges. Denote by $G[S]$ the subgraph of G induced by the vertex set S, i.e., the graph having vertex set S and whose edge set consists of those edges of G incident with two elements of S. Let S and T be two disjoint subsets of $V(G)$, we denote by $E_G(S,T)$ the set of edges with one end in S and the other end in T, and we write

$$e_G(S,T) = |E_G(S,T)|.$$

Let r be a real number. Recall that $\lfloor r \rfloor$ is the greatest integer such that $\lfloor r \rfloor \leq r$. Other definitions and terminologies can be found in [3].

Let g and f be two nonnegative integer-valued functions defined on $V(G)$ such that $g(x) \leq f(x)$ for every $x \in V(G)$. A spanning subgraph F of G satisfying $g(x) \leq d_F(x) \leq f(x)$ for every $x \in V(G)$ is a (g, f)-factor of G. Let $h: E(G) \to [0, 1]$ be a function defined on $E(G)$. If

$$g(x) \leq \sum_{e \ni x} h(e) \leq f(x)$$

holds for every $x \in V(G)$, then we call $G[F_h]$ a **fractional** (g, f)-**factor** of G with indicator function h, where

$$F_h = \{e : e \in E(G), h(e) > 0\}.$$

A fractional (f, f)-factor is called simply a **fractional** f-**factor**. If $f(x) = k$, then a fractional f-factor is called a **fractional** k-**factor**. A graph G is called a **fractional** (g, f, m)-**deleted graph** if there exists a fractional (g, f)-factor $G[F_h]$ of G with indicator function h such that $h(e) = 0$ for any $e \in E(H)$, where H is any subgraph of G with m edges. A fractional (f, f, m)-deleted graph is simply called a **fractional** (f, m)-**deleted graph**. If $f(x) = k$ for any $x \in V(G)$, then a fractional (f, m)-deleted graph is called a **fractional** (k, m)-**deleted graph**. Set $m = 1$. Then a fractional (g, f, m)-deleted graph is a fractional (g, f)-deleted graph; a fractional (f, m)-deleted graph is simple called a **fractional** f-**deleted graph**; a fractional (k, m)-deleted graph is a fractional k-deleted graph. The basic results on graph factors can be found in [1, 17, 24].

1.2 Graphic Parameters

The graphic parameters play important roles in the research of graph factors and fractional factors, they are used frequently as sufficient conditions for the existence of graph factors and fractional factors. Since verifying graphic parameter conditions are often easier than that of characterizations, and as

1.2 Graphic Parameters

well the parameter conditions reflect the structures and properties of graphs from different perspectives, it is quite common in graph theory to investigate the links among the parameters. In this section, we show some graphic parameters, such as binding number, toughness, isolated toughness, independence number, connectivity and minimum degree, etc.

The minimum degree of G is denoted by $\delta(G)$, i.e.,
$$\delta(G) = \min\{d_G(x) : x \in V(G)\}.$$
The maximum degree of G is denoted by $\Delta(G)$, i.e.,
$$\Delta(G) = \max\{d_G(x) : x \in V(G)\}.$$

A vertex subset S of G is called **independent** if $G[S]$ has no edges. An independent set S of G is called a **maximum independent set** if G excludes a independent set S' with $|S'| > |S|$. The number of vertices in the maximum independent set S of G is called the **independence number**, and is denoted by $\alpha(G)$.

Let G be a connected graph.
$$\kappa(G) = \min\{|T| : T \subseteq V(G), G - T \text{ is disconnected}$$
$$\text{or is a trivial graph}\}$$
is called the **connectivity** of G.
$$\lambda(G) = \min\{|E_G(S, V(G) \setminus S)| : S \subseteq V(G)\}$$
is called the **edge-connectivity** of G.

The binding number was introduced by Woodall[21] and is defined as
$$\text{bind}(G) = \min\left\{\frac{|N_G(X)|}{|X|} : \emptyset \neq X \subseteq V(G), N_G(X) \neq V(G)\right\}.$$
Obviously, for any nonempty subset $S \subseteq V(G)$ with $N_G(S) \neq V(G)$, we have
$$|N_G(S)| \geq \text{bind}(G)|S|.$$

The number of connected components in G is denoted by $\omega(G)$. The toughness $t(G)$ of a connected graph G was first defined by Chvatal in [5] as follows.

$$t(G) = \min\left\{\frac{|S|}{\omega(G-S)} : S \subseteq V(G), \omega(G-S) \geq 2\right\},$$

if G is not complete; otherwise, $t(G) = +\infty$.

Enomoto[6] introduced a new parameter $\tau(G)$ which is a slight variation of toughness, but seems better to fit the research of graph factors and fractional factors. For a connected graph G, we define

$$\tau(G) = \min\left\{\frac{|S|}{\omega(G-S) - 1} : S \subseteq V(G), \omega(G-S) \geq 2\right\},$$

if G is not complete; otherwise, $\tau(G) = +\infty$.

we use $i(G)$ to denote the number of isolated vertices of G. The isolated toughness $I(G)$ of a graph G is defined by Ma and Liu[22] as follows.

$$I(G) = \min\left\{\frac{|S|}{i(G-S)} : S \subseteq V(G), i(G-S) \geq 2\right\},$$

if G is not complete; otherwise, $I(G) = +\infty$.

Ma and Liu[15] introduced a new parameter $I'(G)$ which is a slight variation of isolated toughness. For a graph G, we define

$$I'(G) = \min\left\{\frac{|S|}{i(G-S) - 1} : S \subseteq V(G), i(G-S) \geq 2\right\},$$

if G is not complete; otherwise, $I'(G) = +\infty$.

Chapter 2
Fractional Factors

In this chapter we study fractional k-factors and fractional (g, f)-factors which are natural generalizations of k-factors and (g, f)-factors, and investigate fractional k-factors and fractional (g, f)-factors including any given edges. This chapter is divided into three parts. Firstly, We show some sufficient conditions for graphs to have fractional k-factors. Secondly, we give some results on fractional k-factors including any given edges. Thirdly, we obtain some sufficient conditions for graphs to have fractional (g, f)-factors with prescribed properties.

2.1 Fractional k-Factors

Let k be an integer such that $k \geq 1$. Then a spanning subgraph F of G is called a k-**factor** if $d_F(x) = k$ for all $x \in V(G)$. Let $h : E(G) \to [0, 1]$ be a function. If $\sum_{e \ni x} h(e) = k$ holds for each $x \in V(G)$, then we call $G[F_h]$ a **fractional k-factor** of G with indicator function h where

$$F_h = \{e \in E(G) : h(e) > 0\}.$$

A fractional 1-factor is also called a **fractional perfect matching**. If $h(e) \in \{0, 1\}$ for any $e \in E(G)$, then a fractional k-factor is equivalent to a k-factor, and so a fractional k-factor is a natural generalization of a k-factor.

We shall show a necessary and sufficient condition for a graph to have a fractional k-factor, which is a special case of the fractional (g, f)-factor theorem presented by Anstee[2]. Liu and Zhang[12] showed a simple proof of the theorem.

Theorem 2.1.1 [2, 12] *Let G be a graph. Then G has a fractional k-factor if and only if for every subset S of $V(G)$,*

$$\delta_G(S,T) = k|S| - k|T| + d_{G-S}(T) \geq 0,$$

where $T = \{x : x \in V(G) \setminus S, d_{G-S}(x) \leq k\}$.

The following theorem obtained by Zhang and Liu[25] is equivalent to Theorem 2.1.1.

Theorem 2.1.2 [25] *Let G be a graph. Then G has a fractional k-factor if and only if for all subset S of $V(G)$,*

$$k|S| - \sum_{j=0}^{k-1}(k-j)p_j(G-S) \geq 0,$$

where $p_j(G-S)$ denotes the number of vertices in $G-S$ with degree j.

In Theorem 2.1.2, let $k = 1$, then we get the following result.

Theorem 2.1.3 [18] *A graph G has a fractional 1-factor if and only if*

$$i(G-S) \leq |S|$$

for all $S \subseteq V(G)$.

In [7], Iida and Nishimura gave a neighborhood condition for a graph to have a k-factor.

Theorem 2.1.4 [7] *Let k be an integer such that $k \geq 2$, and let G be a connected graph of order n such that*

$$n \geq 9k - 1 - 4\sqrt{2(k-1)^2 + 2},$$

kn is even, and the minimum degree is at least k. If G satisfies

$$|N_G(x) \cup N_G(y)| \geq \frac{1}{2}(n+k-2)$$

for each pair of nonadjacent vertices $x, y \in V(G)$, then G has a k-factor.

2.1 Fractional k-Factors

In the following we shall present some results on fractional k-factors by using Theorems 2.1.1–2.1.3.

We first give some neighborhood conditions for the existence of fractional k-factors in graphs.[37, 40, 42]

Theorem 2.1.5 [40] Let G be a connected graph of order n such that $n \geq 3$. If
$$|N_G(x) \cup N_G(y)| \geq \frac{n}{2}$$
for each pair of nonadjacent vertices $x, y \in V(G)$, then G has a fractional 1-factor.

Theorem 2.1.6 [40] Let k be an integer such that $k \geq 1$, and let G be a connected graph of order n such that
$$n \geq 9k - 1 - 4\sqrt{2(k-1)^2 + 2},$$
and the minimum degree $\delta(G) \geq k$. If
$$|N_G(x) \cup N_G(y)| \geq \max\left\{\frac{n}{2}, \frac{1}{2}(n + k - 2)\right\}$$
for each pair of nonadjacent vertices $x, y \in V(G)$, then G has a fractional k-factor.

The following lemmas is often applied in the proof of Theorems 2.1.5–2.1.6.

Lemma 2.1.1 [7] Let k be an integer such that $k \geq 1$. Then
$$9k - 1 - 4\sqrt{2(k-1)^2 + 2} \begin{cases} > 3k + 5, & \text{for } k \geq 4, \\ > 3k + 4, & \text{for } k = 3, \\ = 3k + 3, & \text{for } k = 2, \\ > 2, & \text{for } k = 1. \end{cases}$$

Lemma 2.1.2 [40] Let G be a connected graph of order n. If
$$|N_G(x) \cup N_G(y)| \geq \frac{n}{2}$$

for each pair of nonadjacent vertices $x, y \in V(G)$, then
$$\omega(G - S) \leq |S| + 1$$
for all $S \subseteq V(G)$ with $|S| \geq 2$.

Proof Suppose that there exists a vertex subset $S \subseteq V(G)$ with $|S| \geq 2$ such that $\omega(G - S) \geq |S| + 2$.

Claim $2 \leq |S| \leq \dfrac{n-2}{2}$.

Proof If $|S| \geq \frac{n-1}{2}$, then
$$\omega(G - S) \leq n - |S| \leq n - \frac{n-1}{2} = \frac{n+1}{2} = \frac{n-1}{2} + 1$$
$$\leq |S| + 1,$$
a contradiction. □

In the following, let $C_1, C_2, \cdots, C_\omega$ be the connected components of $G - S$. We have
$$\omega = \omega(G - S) \geq |S| + 2 \geq 4$$
since $|S| \geq 2$. Choose an arbitrary vertex $x_i \in V(C_i)$ ($1 \leq i \leq \omega(G - S)$). Then $x_i x_j \notin E(G)$ ($i \neq j$). By the hypothesis of the lemma, then
$$\frac{n}{2} \leq |N_G(x_i) \cup N_G(x_j)| \leq d_{G-S}(x_i) + d_{G-S}(x_j) + |S|$$
$$\leq |V(C_i)| - 1 + |V(C_j)| - 1 + |S|,$$
that is,
$$|V(C_i)| + |V(C_j)| \geq \frac{n+4}{2} - |S|$$
for $i \neq j$. Thus, we get that
$$n = |S| + |V(C_1)| + |V(C_2)| + \cdots + |V(C_\omega)|$$
$$= |S| + \frac{1}{2}[2(|V(C_1)| + |V(C_2)| + \cdots + |V(C_\omega)|)]$$
$$\geq |S| + \frac{\omega}{2}\left(\frac{n+4}{2} - |S|\right)$$
$$\geq |S| + \frac{|S| + 2}{2}\left(\frac{n+4}{2} - |S|\right)$$

2.1 Fractional k-Factors

$$= -\frac{1}{2}|S|^2 + \frac{n+4}{4}|S| + \frac{n+4}{2},$$

that is,

$$n \geq -\frac{1}{2}|S|^2 + \frac{n+4}{4}|S| + \frac{n+4}{2}. \tag{2.1.1}$$

Let

$$f(|S|) = -\frac{1}{2}|S|^2 + \frac{n+4}{4}|S| + \frac{n+4}{2}.$$

In fact, the function $f(|S|)$ attains its minimum value at $|S| = 2$ or $|S| = \frac{n-2}{2}$ which follows from $2 \leq |S| \leq \frac{n-2}{2}$, and the minimum value of the function $f(|S|)$ is $\min\{n+2, \frac{5n+2}{4}\}$. According to (2.1.1) and the minimum value of the function $f(|S|)$, we have

$$n > n,$$

which is a contradiction. Completing the proof of the lemma. □

Proof of Theorem 2.1.5 If G is a graph which satisfies the condition of the theorem and there is no fractional 1-factor in G, then by Theorem 2.1.3, there exists a vertex set $S \subseteq V(G)$ such that

$$i(G - S) > |S|. \tag{2.1.2}$$

Clearly, $S \neq \emptyset$ since G is connected. In the following, we assume $|S| \geq 1$. The proof splits into two cases.

Case 1 $|S| = 1$.

Thus, we get that $i(G - S) > |S| = 1$, that is,

$$i(G - S) \geq 2.$$

Let $x, y \in I(G - S)$, then $xy \notin E(G)$ and $d_{G-S}(x) = d_{G-S}(y) = 0$. According to the hypothesis of the theorem, we have

$$\frac{n}{2} \leq |N_G(x) \cup N_G(y)| \leq d_{G-S}(x) + d_{G-S}(y) + |S| = |S| = 1.$$

Therefore, we get $n \leq 2$. This contradicts $n \geq 3$.

Case 2 $|S| \geq 2$.

We have known that $\omega(G - S) \geq i(G - S)$ and it implies that

$$\omega(G - S) \geq |S| + 1$$

by (2.1.2). But $\omega(G-S) \leq |S|+1$ for all $S \subseteq V(G)$ with $|S| \geq 2$ by Lemma 2.1.2. Thus, we have
$$\omega(G-S) = |S|+1$$
for all $S \subseteq V(G)$ with $|S| \geq 2$. Moreover, we have
$$i(G-S) = |S|+1$$
and
$$|S| \leq \frac{n-1}{2}. \tag{2.1.3}$$

For any two vertices $x, y \in I(G-S)$, obviously, $xy \notin E(G)$ and $d_{G-S}(x) = d_{G-S}(y) = 0$. Then, we have
$$\frac{n}{2} \leq |N_G(x) \cup N_G(y)| \leq d_{G-S}(x) + d_{G-S}(y) + |S| = |S| \leq \frac{n-1}{2}$$
by the hypothesis of the theorem and (2.1.3). It is a contradiction. Completing the proof of the theorem. □

Proof of Theorem 2.1.6 By Theorem 2.1.5, the result obviously holds for $k = 1$. If $k \geq 2$ and kn is even, then G has a k-factor by Theorem 2.1.4. We have known that a k-factor is a special fractional k-factor. Now we consider the case that k and n are both odd.

If G has no fractional k-factor, then by Theorem 2.1.1, there exists some $S \subseteq V(G)$ such that
$$\delta_G(S,T) = k|S| + d_{G-S}(T) - k|T| \leq -1, \tag{2.1.4}$$
where
$$T = \{x : x \in V(G) \setminus S, d_{G-S}(x) \leq k\}.$$

We choose subsets S and T such that $|T|$ is minimum and S and T satisfy (2.1.4). As k is odd, therefore, $(k-1)$ is even and G has a fractional $(k-1)$-factor by Theorem 2.1.4.

Claim 1 $|T| \geq |S|+1$.

Proof Since G has a fractional $(k-1)$-factor, then we get
$$(k-1)|S| + d_{G-S}(T') - (k-1)|T'| \geq 0$$
where $T' = \{x : x \in V(G) \setminus S, d_{G-S}(x) \leq k-1\}$. Moreover,

2.1 Fractional k-Factors

$$\begin{aligned}
0 &\leq (k-1)|S| + d_{G-S}(T') - (k-1)|T'| \\
&= (k-1)|S| + d_{G-S}(T) - kp_k(G-S) \\
&\quad - (k-1)(|T| - p_k(G-S)) \\
&= k|S| + d_{G-S}(T) - k|T| - |S| + |T| - p_k(G-S) \\
&= \delta_G(S,T) - |S| + |T| - p_k(G-S) \\
&\leq -1 - |S| + |T| - p_k(G-S),
\end{aligned}$$

where $p_k(G-S)$ denotes the number of vertices in $G-S$ with degree k.

Thus we may obtain,

$$|T| \geq |S| + 1.$$

Claim 1 is proved. □

Claim 2 $|T| \geq k+1$.

Proof If $|T| \leq k$, then by (2.1.4) and

$$|S| + d_{G-S}(x) \geq d_G(x) \geq \delta(G) \geq k,$$

we obtain

$$\begin{aligned}
-1 &\geq \delta_G(S,T) = k|S| + d_{G-S}(T) - k|T| \\
&\geq |T||S| + d_{G-S}(T) - k|T| \\
&= \sum_{x \in T}(|S| + d_{G-S}(x) - k) \geq 0,
\end{aligned}$$

which is a contradiction. □

Claim 3 $d_{G-S}(x) \leq k-1$ for all $x \in T$.

Proof If $d_{G-S}(x) \geq k$ for some $x \in T$, then the subsets S and $T \setminus \{x\}$ satisfy (2.1.4). This contradicts the choice of S and T. □

Claim 4 $|S| \leq \dfrac{n-1}{2}$.

Proof By Claim 1 and $n \geq |S| + |T|$, we obtain

$$n \geq |S| + |T| \geq 2|S| + 1,$$

which implies $|S| \leq \frac{n-1}{2}$. □

Define $h_1 = \min\{d_{G-S}(x) | x \in T\}$, and

$$T' = \{x \in T, d_{G-S}(x) = 0\},$$

and $p = |T'|$. Then by Claim 3 and

$$|S| + d_{G-S}(x) \geq d_G(x) \geq \delta(G) \geq k,$$

we have

$$h_1 \leq k-1, \quad k \leq \delta(G) \leq h_1 + |S|. \tag{2.1.5}$$

Choose $x_1 \in T$ such that $d_{G-S}(x_1) = h_1$. If $T \setminus N_T[x_1] \neq \emptyset$, let

$$h_2 = \min\{d_{G-S}(x) | x \in T \setminus N_T[x_1]\}.$$

Thus, we have $0 \leq h_1 \leq h_2 \leq k-1$ by Claim 3.

We shall consider various cases according to the value of p and derive contradictions.

Case 1 $p \geq 2$.

Then we have

$$i(G-S) = |I(G-S)| \geq 2.$$

Thus, there exist $x, y \in I(G-S)$ and $d_{G-S}(x) = d_{G-S}(y) = 0$ and $xy \notin E(G)$. By the hypothesis of the theorem and Claim 4 and $k \geq 3$, we get that

$$\frac{n+k-2}{2} \leq |N_G(x) \cup N_G(y)| \leq d_{G-S}(x) + d_{G-S}(y) + |S|$$

$$= |S| \leq \frac{n-1}{2},$$

a contradiction.

Case 2 $p = 1$.

Obviously, we have $h_1 = 0$ and $|N_T[x_1]| = 1$. Thus, we get that $T \setminus N_T[x_1] \neq \emptyset$ and $1 \leq h_2 \leq k-1$ in view of Claim 2 and $p = 1$. Choose $x_2 \in T \setminus N_T[x_1]$ such that $d_{G-S}(x_2) = h_2$. Clearly, $x_1 x_2 \notin E(G)$. By the hypothesis of the theorem, we obtain

$$\frac{n+k-2}{2} \leq |N_G(x_1) \cup N_G(x_2)|$$

$$\leq d_{G-S}(x_1) + d_{G-S}(x_2) + |S|$$

$$= h_2 + |S|,$$

that is,
$$|S| \geq \frac{n+k-2}{2} - h_2. \qquad (2.1.6)$$

According to Lemma 2.1.1 and $k \geq 3$, we get $n > 3k+4$. By (2.1.4), (2.1.6) and $n \geq |S| + |T|$, we have

$$\begin{aligned}
-1 \geq \delta_G(S,T) &= k|S| + d_{G-S}(T) - k|T| \\
&\geq k|S| + h_2(|T|-1) - k|T| \\
&= k|S| - (k-h_2)|T| - h_2 \\
&\geq k|S| - (k-h_2)(n-|S|) - h_2 \\
&= (2k-h_2)|S| + (h_2-k)n - h_2 \\
&\geq (2k-h_2)\left(\frac{n+k-2}{2} - h_2\right) + (h_2-k)n - h_2 \\
&= h_2^2 + \frac{n-5k}{2}h_2 + k^2 - 2k \\
&> h_2^2 + \frac{3k+4-5k}{2}h_2 + k^2 - 2k \\
&= h_2^2 + 2h_2 + k(k-h_2-2).
\end{aligned}$$

If $1 \leq h_2 \leq k-2$, then
$$-1 \geq \delta_G(S,T) = k|S| + d_{G-S}(T) - k|T| \geq h_2^2 + 2h_2 > 0,$$
a contradiction.

If $h_2 = k-1$, then
$$\begin{aligned}
-1 \geq \delta_G(S,T) &= k|S| + d_{G-S}(T) - k|T| \\
&\geq h_2^2 + 2h_2 + k(k-h_2-2) \\
&= k^2 - k - 1 > 0.
\end{aligned}$$

It is a contradiction.

Case 3 $p = 0$.

Claim 5 $T \setminus N_T[x_1] \neq \emptyset$.

Proof If $T = N_T[x_1]$, then we have
$$|T| = |N_T[x_1]| \leq d_{G-S}(x_1) + 1 = h_1 + 1 \leq k,$$

which contradicts Claim 2.

According to Claim 5, there exists $x_2 \in T \setminus N_T[x_1]$ such that $d_{G-S}(x_2) = h_2$. Clearly, $x_1x_2 \notin E(G)$. In view of the hypothesis of the theorem, we have

$$\frac{n+k-2}{2} \leq |N_G(x_1) \cup N_G(x_2)|$$
$$\leq d_{G-S}(x_1) + d_{G-S}(x_2) + |S|$$
$$= h_1 + h_2 + |S|,$$

that is,
$$2|S| \geq n + k - 2(h_1 + h_2 + 1). \qquad (2.1.7)$$

Since $p = 0$, then we have $1 \leq h_1 \leq h_2 \leq k-1$. Moreover,
$$|N_T[x_1]| \leq d_{G-S}(x_1) + 1 = h_1 + 1.$$

Obviously, $n - |S| - |T| \geq 0$ and $k - h_2 \geq 1$. Thus,

$$(k - h_2)(n - |S| - |T|) - 1$$
$$\geq \delta_G(S,T) = k|S| + d_{G-S}(T) - k|T|$$
$$\geq k|S| - k|T| + h_1|N_T[x_1]| + h_2(|T| - |N_T[x_1]|).$$

Since $|N_T[x_1]| \leq h_1 + 1$, this implies
$$(2k - h_2)|S| \leq (k - h_2)n + (h_2 - h_1)(h_1 + 1) - 1. \qquad (2.1.8)$$

By (2.1.7), (2.1.8), $1 \leq h_1 \leq h_2 \leq k-1$ and
$$n \geq 9k - 1 - 4\sqrt{2(k-1)^2 + 2},$$

we have
$$0 \leq -h_2 n + 2(h_2 - h_1)(h_1 + 1) - 2 - k(2k - h_2)$$
$$+ 2(2k - h_2)(h_1 + h_2 + 1)$$
$$\leq -4h_2^2 + (9k - n - 2)h_2 - (2k^2 - 4k + 2)$$
$$\leq -4h_2^2 + (4\sqrt{2(k-1)^2 + 2} - 1)h_2 - 2(k-1)^2$$
$$= -4\left(h_2 - \sqrt{\frac{(k-1)^2 + 1}{2}}\right)^2 - h_2 + 2,$$

that is,
$$0 \leq -4\left(h_2 - \sqrt{\frac{(k-1)^2+1}{2}}\right)^2 - h_2 + 2. \qquad (2.1.9)$$

Subcase 3.1 $3 \leq h_2 \leq k-1$.

In view of (2.1.9), we get
$$0 \leq -4\left(h_2 - \sqrt{\frac{(k-1)^2+1}{2}}\right)^2 - h_2 + 2 \leq -1.$$

This is a contradiction.

Subcase 3.2 $h_2 = 2$.

According to (2.1.9) and $k \geq 3$ (k is an odd integer), we obtain
$$0 \leq -4\left(2 - \sqrt{\frac{(k-1)^2+1}{2}}\right)^2 < 0.$$

This is a contradiction.

Subcase 3.3 $h_2 = 1$.

By (2.1.9) and $k \geq 3$ (k is an odd integer), we have
$$0 \leq -4\left(1 - \sqrt{\frac{(k-1)^2+1}{2}}\right)^2 + 1 < 0.$$

This is a contradiction.

From all the cases above, we deduced the contradiction. Hence, G has a fractional k-factor. Completing the proof of the theorem. □

Woodall[20] obtained a sufficient condition for a graph to have a k-factor by using neighborhoods of independent sets.

Theorem 2.1.7 [20] Let $k \geq 2$ be an integer and G a graph of order n with $n \geq 4k-6$, where if k is odd, then n is even and G is connected. Let G satisfy
$$|N_G(X)| \geq \frac{(k-1)n + |X| - 1}{2k-1} \qquad (2.1.10)$$
for every non-empty independent subset X of $V(G)$, and
$$\delta(G) \geq \frac{k-1}{2k-1}(n+2). \qquad (2.1.11)$$
Then G has a k-factor.

we now discuss a sufficient condition on neighborhoods of independent sets for a graph to have a fractional k-factor, which is an extension of Theorem 2.1.7.

Theorem 2.1.8 [37] *Let k be a positive integer and G a graph of order n with $n \geq 4k - 6$. Then*

(1) if k is even and (2.1.10) and (2.1.11) hold, then G has a fractional k-factor; and

(2) if k is odd, and

$$|N_G(X)| > \frac{(k-1)n + |X| - 1}{2k - 1}$$

for every non-empty independent subset X of $V(G)$, and

$$\delta(G) > \frac{k-1}{2k-1}(n+2),$$

then G has a fractional k-factor.

Proof If k is even and (2.1.10) and (2.1.11) hold, then G has a k-factor by Theorem 2.1.7. But a k-factor is a special type of fractional k-factor, and so G has a fractional k-factor. In the following, we assume that k is odd.

Case 1 $k = 1$.

For any $S \subseteq V(G)$, we denote by Y the set of isolated vertices in $G - S$, so that $|Y| = i(G - S)$. If $Y = \emptyset$, that is, $i(G - S) = 0$, then obviously $i(G - S) \leq |S|$. In the following we assume $Y \neq \emptyset$. According to the condition of the theorem, we have

$$|S| \geq |N_G(Y)| > |Y| - 1 = i(G - S) - 1.$$

In view of the integrity of $|S|$ and $i(G - S)$, we get

$$i(G - S) \leq |S|.$$

By Theorem 2.1.3, G has a fractional 1-factor.

Case 2 $k \geq 3$ is odd.

Let G be a graph satisfying the hypotheses of Theorem 2.1.8, which has no fractional k-factor. Then, by Theorem 2.1.1, there exist disjoint subsets

S, T of $V(G)$ such that
$$\delta_G(S,T) = k|S| + d_{G-S}(T) - k|T| \leq -1. \tag{2.1.12}$$
We choose disjoint subsets S and T satisfying (2.1.12) such that $|T|$ is as small as possible.

Claim 1 $|T| \geq k+1$.

Proof Note first that
$$2k^2 - 7k + 4 > 2k^2 - 8k + 6 = 2(k-1)(k-3) \geq 0,$$
so that, by the hypotheses of the theorem,
$$\delta(G) > \frac{(k-1)(n+2)}{2k-1} \geq \frac{(k-1)(4k-4)}{2k-1}$$
$$= \frac{2k^2 - 7k + 4}{2k-1} + k > k.$$

So if $|T| \leq k$ then, by (2.1.12),
$$-1 \geq \delta_G(S,T) = k|S| + d_{G-S}(T) - k|T|$$
$$\geq |T||S| + d_{G-S}(T) - k|T|$$
$$= \sum_{x \in T}(|S| + d_{G-S}(x) - k)$$
$$\geq \sum_{x \in T}(\delta(G) - k) > 0.$$

This is a contradiction. □

Claim 2 $d_{G-S}(x) \leq k-1$ for all $x \in T$.

Proof If $d_{G-S}(x) \geq k$ for some $x \in T$, then the subsets S and $T \setminus \{x\}$ satisfy (2.1.12). This contradicts the choice of S and T. □

According to Claim 1, we have $T \neq \emptyset$. Hence, we may define
$$h = \min\{d_{G-S}(x) : x \in T\}.$$
Clearly, we have
$$\delta(G) \leq h + |S|. \tag{2.1.13}$$
In view of Claim 2, we obtain
$$0 \leq h \leq k - 1.$$

Subcase 2.1 $h = 0$.

Let $Y = \{x \in T : d_{G-S}(x) = 0\}$. Obviously, $Y \neq \emptyset$ and Y is independent. Thus, we obtain by the condition of the theorem

$$\frac{(k-1)n + |Y| - 1}{2k - 1} < |N_G(Y)| \leq |S|. \qquad (2.1.14)$$

According to (2.1.12), (2.1.14) and $|S| + |T| \leq n$, we have

$$-1 \geq \delta_G(S,T) = k|S| + d_{G-S}(T) - k|T|$$
$$\geq k|S| + |T| - |Y| - k|T|$$
$$= k|S| - (k-1)|T| - |Y|$$
$$\geq k|S| - (k-1)(n - |S|) - |Y|$$
$$= (2k-1)|S| - |Y| - (k-1)n$$
$$> (2k-1)\left(\frac{(k-1)n + |Y| - 1}{2k-1}\right) - |Y| - (k-1)n$$
$$= -1.$$

It is a contradiction.

Subcase 2.2 $1 \leq h \leq k-1$.

By (2.1.12) and $|S| + |T| \leq n$ and $k - h \geq 1$, we get that

$$-1 \geq \delta_G(S,T) = k|S| + d_{G-S}(T) - k|T|$$
$$\geq k|S| - (k-h)|T|$$
$$\geq k|S| - (k-h)(n - |S|)$$
$$= (2k-h)|S| - (k-h)n.$$

This inequality implies

$$|S| \leq \frac{(k-h)n - 1}{2k - h}. \qquad (2.1.15)$$

By a hypothesis of the theorem, and using (2.1.13) and (2.1.15),

$$\frac{(k-1)(n+2) + 1}{2k - 1} \leq \delta(G) \leq |S| + h \leq \frac{(k-h)n - 1}{2k - h} + h. \qquad (2.1.16)$$

If the LHS and RHS of (2.1.16) are denoted by A and B respectively, then (2.1.16) says that $A - B \leq 0$. But, after some rearranging, we find that

2.1 Fractional k-Factors

$$(2k-1)(2k-h)(A-B) = (h-1)[kn-(2k-h)(2k-1)] + 2k-1. \tag{2.1.17}$$

Clearly this is positive if $h = 1$. Recall that $n \geq 4k - 6$ by a hypothesis of the theorem. It follows that if $h \geq 2$ then

$$kn > (2k-h)(2k-1),$$

unless $n = 4k - 6$ and $h = 2$, when $kn = (2k-h)(2k-1) - 2$. Since $k \geq 3$, it is clear that the expression in (2.1.17) is positive in all cases, and this contradicts (2.1.16).

From the contradictions we deduce that G has a fractional k-factor. This completes the proof of Theorem 2.1.8. □

Remark 2.1.1 Let k be odd. Then, let us show that the condition

$$|N_G(X)| > \frac{(k-1)n + |X| - 1}{2k-1}$$

in Theorem 2.1.8 can not be replaced by

$$|N_G(X)| \geq \frac{(k-1)n + |X| - 1}{2k-1}.$$

Let t be an odd positive integer with $t \geq 3$. Let

$$H = K_{(k-1)t+1} \vee \left((2kK_1) \cup \left(\frac{(t-2)k-1}{2} K_2\right)\right),$$

where $\vee$ denotes "join". Then H has order $n = (2k-1)t$. Let $X = V(2kK_1)$. Then

$$\delta(H) = (k-1)t + 1 = \frac{(2k-1)(k-1)t + 2k - 1}{2k-1}$$

$$= \frac{(k-1)n + 2k - 1}{2k-1} = \frac{(k-1)n + |X| - 1}{2k-1}$$

$$> \frac{(k-1)n + 2k - 2}{2k-1} = \frac{k-1}{2k-1}(n+2)$$

and

$$|N_H(X)| = (k-1)t + 1 = \frac{(k-1)n + |X| - 1}{2k-1},$$

and it is easy to see from this that

$$|N_H(X)| \geq \frac{(k-1)n + |X| - 1}{2k - 1}$$

for every non-empty independent subset X of $V(H)$. Let

$$S = V(K_{(k-1)t+1}) \subseteq V(H),$$

$$T = V((2kK_1) \cup \left(\frac{(t-2)k-1}{2}K_2\right)) \subseteq V(H).$$

Then $|S| = (k-1)t + 1$, $|T| = kt - 1$, and $d_{H-S}(T) = k(t-2) - 1$. Thus, we obtain

$$\begin{aligned}\delta_H(S,T) &= k|S| + d_{H-S}(T) - k|T| \\ &= k((k-1)t+1) + k(t-2) - 1 - k(kt-1) \\ &= -1 < 0.\end{aligned}$$

By Theorem 2.1.1, there are no fractional k-factors in H. In the above sense, the condition

$$|N_G(X)| > \frac{(k-1)n + |X| - 1}{2k - 1}$$

in Theorem 2.1.8 is best possible.

Let k be even. Then, let us show that the condition

$$|N_G(X)| \geq \frac{(k-1)n + |X| - 1}{2k - 1}$$

in Theorem 2.1.8 can not be replaced by

$$|N_G(X)| \geq \frac{(k-1)n + |X| - 2}{2k - 1}.$$

Let $t \geq 4$ be an integer, and

$$H = K_{(k-1)t+1} \vee \left((2k+1)K_1 \cup \left(\frac{k(t-2)-2}{2}K_2\right)\right).$$

Then H has order $n = (2k-1)t$. Let $X = V((2k+1)K_1)$. Then

$$\delta(H) = (k-1)t + 1 = \frac{(2k-1)(k-1)t + 2k - 1}{2k - 1}$$

$$= \frac{(k-1)n + 2k - 1}{2k - 1} = \frac{(k-1)n + |X| - 2}{2k - 1}$$

2.1 Fractional k-Factors

$$> \frac{(k-1)n + 2k - 2}{2k-1} = \frac{k-1}{2k-1}(n+2)$$

and

$$|N_H(X)| = (k-1)t + 1 = \frac{(k-1)n + |X| - 2}{2k-1},$$

and it is easy to see from this that

$$|N_H(X)| \geq \frac{(k-1)n + |X| - 2}{2k-1}$$

for every non-empty independent subset X of $V(H)$. Let

$$S = V(K_{(k-1)t+1}) \subseteq V(H),$$

$$T = V\left((2k+1)K_1 \cup \left(\frac{k(t-2)-2}{2}K_2\right)\right) \subseteq V(H).$$

Then $|S| = (k-1)t+1$, $|T| = kt-1$, and $d_{H-S}(T) = k(t-2) - 2$. Thus, we get

$$\delta_H(S,T) = k|S| + d_{H-S}(T) - k|T|$$
$$= k((k-1)t+1) + k(t-2) - 2 - k(kt-1)$$
$$= -2 < 0.$$

By Theorem 2.1.1, there are no fractional k-factors in H. In the above sense, the result in Theorem 2.1.8 is best possible.

The following result on fractional k-factors was obtained by Zhou, Pu and Xu[42].

Theorem 2.1.9 [42] Let k be an integer with $k \geq 1$, and let G be a graph of order n with $n \geq 6k - 12 + \frac{6}{k}$. Suppose for any subset $X \subset V(G)$, we have

$$N_G(X) = V(G) \quad \text{if} \quad |X| \geq \left\lfloor \frac{kn}{2k-1} \right\rfloor;$$

or

$$|N_G(X)| \geq \frac{2k-1}{k}|X| \quad \text{if} \quad |X| < \left\lfloor \frac{kn}{2k-1} \right\rfloor.$$

Then G has a fractional k-factor.

Lemma 2.1.3 [42] *Let G be a graph of order n which satisfies the assumption of Theorem 2.1.9. Then*

$$\delta(G) \geq \frac{(k-1)n+k}{2k-1}.$$

Proof Let x be a vertex of G with degree $\delta(G)$. Set $X = V(G) \setminus N_G(x)$. Obviously, $x \notin N_G(X)$ and $N_G(X) \neq V(G)$. Thus, we obtain

$$n - 1 \geq |N_G(X)| \geq \frac{2k-1}{k}|X|,$$

that is,

$$(2k-1)|X| \leq k(n-1). \tag{2.1.18}$$

Using (2.1.18) and $|X| = n - \delta(G)$, we have

$$(2k-1)(n - \delta(G)) \leq k(n-1).$$

Hence

$$\delta(G) \geq \frac{(k-1)n+k}{2k-1}.$$

This completes the proof of Lemma 2.1.3. □

Proof of Theorem 2.1.9 Let G be a graph satisfying the hypotheses of Theorem 2.1.9, which has no fractional k-factor. Then by Lemma 2.1.1, there exists some $S \subseteq V(G)$ such that

$$\delta_G(S, T) = k|S| + d_{G-S}(T) - k|T| \leq -1. \tag{2.1.19}$$

where $T = \{x : x \in V(G) \setminus S, d_{G-S}(x) \leq k - 1\}$. Obviously, $T \neq \emptyset$ by (2.1.19). Define

$$h = \min\{d_{G-S}(t) : t \in T\}.$$

From the definition of T, we obtain

$$0 \leq h \leq k - 1.$$

Case 1 $2 \leq h \leq k - 1$.

In terms of Lemma 2.1.3 and the definition of h, we get

$$|S| + h \geq \delta(G) \geq \frac{(k-1)n+k}{2k-1}. \tag{2.1.20}$$

According to (2.1.19) and $|S| + |T| \leq n$, we obtain

2.1 Fractional k-Factors

$$\begin{aligned}
-1 \geq \delta_G(S,T) &= k|S| + d_{G-S}(T) - k|T| \\
&\geq k|S| + h|T| - k|T| = k|S| - (k-h)|T| \\
&\geq k|S| - (k-h)(n - |S|) \\
&= (2k-h)|S| - (k-h)n.
\end{aligned}$$

This inequality implies

$$|S| \leq \frac{(k-h)n - 1}{2k - h}. \tag{2.1.21}$$

From (2.1.20) and (2.1.21), we have

$$\frac{(k-1)n + k}{2k-1} \leq \delta(G) \leq |S| + h \leq \frac{(k-h)n - 1}{2k - h} + h. \tag{2.1.22}$$

If the LHS and RHS of (2.1.22) are denoted by A and B respectively, then (2.1.22) says that $A - B \leq 0$. But, after some rearranging, we find that

$$(2k-1)(2k-h)(A - B)$$
$$= (h-1)(kn - (2k-1)(2k-h) + k - 1)$$
$$\quad - 2k^2 + 5k - 2. \tag{2.1.23}$$

Since $n \geq 6k - 12 + \frac{6}{k}$, we obtain

$$kn - (2k-1)(2k-2) + k - 1 \geq 2k^2 - 5k + 3 \geq 0. \tag{2.1.24}$$

Using (2.1.23), (2.1.24), $2 \leq h \leq k-1$ and $n \geq 6k - 12 + \frac{6}{k}$, we get

$$(2k-1)(2k-h)(A-B)$$
$$= (h-1)(kn - (2k-1)(2k-h) + k - 1) - 2k^2 + 5k - 2$$
$$\geq (h-1)(kn - (2k-1)(2k-2) + k - 1) - 2k^2 + 5k - 2$$
$$\geq kn - (2k-1)(2k-2) + k - 1 - 2k^2 + 5k - 2$$
$$= kn - 6k^2 + 12k - 5 \geq 1.$$

This inequality implies $A - B > 0$, which contradicts $A - B \leq 0$.

Case 2 $h = 1$.

Subcase 2.1 $|T| \geq \left\lfloor \dfrac{kn}{2k-1} \right\rfloor + 1$.

In terms of the definition of h and $h = 1$, there exists $t \in T$ such that

$d_{G-S}(t) = h = 1$. Thus, we obtain

$$t \notin N_G(T \setminus N_G(t)),$$

which implies

$$N_G(T \setminus N_G(t)) \neq V(G). \tag{2.1.25}$$

On the other hand, using $|T| \geq \lfloor \frac{kn}{2k-1} \rfloor + 1$ and $d_{G-S}(t) = 1$, we have

$$|T \setminus N_G(t)| \geq |T| - 1 \geq \left\lfloor \frac{kn}{2k-1} \right\rfloor.$$

Combining the condition of Theorem 2.1.9, this above inequality implies

$$N_G(T \setminus N_G(t)) = V(G).$$

Which contradicts (2.1.25).

Subcase 2.2 $|T| \leq \left\lfloor \frac{kn}{2k-1} \right\rfloor$.

Since $h = 1$, there exists $u \in T$ such that $d_{G-S}(u) = 1$. Thus, from Lemma 2.1.3, we have

$$|S| + 1 = |S| + d_{G-S}(u) \geq d_G(u) \geq \delta(G) \geq \frac{(k-1)n+k}{2k-1},$$

that is,

$$|S| \geq \frac{(k-1)n+k}{2k-1} - 1 = \frac{(k-1)(n-1)}{2k-1}. \tag{2.1.26}$$

Subcase 2.2.1 $|T| > \frac{k(n-1)}{2k-1}$.

In terms of (2.1.26) and $|T| > \frac{k(n-1)}{2k-1}$, we get

$$|S| + |T| > \frac{(k-1)(n-1)}{2k-1} + \frac{k(n-1)}{2k-1} = n - 1.$$

Combining this with $|S| + |T| \leq n$, we obtain

$$|S| + |T| = n. \tag{2.1.27}$$

According to (2.1.19), (2.1.27) and

$$|T| \leq \left\lfloor \frac{kn}{2k-1} \right\rfloor \leq \frac{kn}{2k-1},$$

2.1 Fractional k-Factors

we have

$$-1 \geq \delta_G(S,T) = k|S| + d_{G-S}(T) - k|T|$$
$$\geq k|S| + |T| - k|T| = k|S| - (k-1)|T|$$
$$= k(n - |T|) - (k-1)|T| = kn - (2k-1)|T|$$
$$\geq kn - (2k-1) \cdot \frac{kn}{2k-1} = 0,$$

which is a contradiction.

Subcase 2.2.2 $|T| \leq \frac{k(n-1)}{2k-1}$.

Since $k - 1 \geq h = 1$, we obtain $k \geq 2$ in this case. Set

$$p = |\{t : t \in T, d_{G-S}(t) = 1\}|.$$

Clearly, $|T| \geq p$. Combining this with (2.1.26) and $k \geq 2$ and $|T| \leq \frac{k(n-1)}{2k-1}$, we obtain

$$\delta_G(S,T) = k|S| + d_{G-S}(T) - k|T|$$
$$\geq k|S| + 2(|T| - p) + p - k|T|$$
$$= k|S| - (k-2)|T| - p$$
$$\geq k \cdot \frac{(k-1)(n-1)}{2k-1} - (k-2) \cdot \frac{k(n-1)}{2k-1} - p$$
$$= \frac{k(n-1)}{2k-1} - p \geq |T| - p \geq 0.$$

That contradicts (2.1.19).

Case 3 $h = 0$.

Let m be the number of vertices x in T such that $d_{G-S}(x) = 0$. Clearly, $m \geq 1$ by $h = 0$. Set $Y = V(G) \setminus S$. Then $N_G(Y) \neq V(G)$ since $h = 0$.

Claim 1 $|Y| < \lfloor \frac{kn}{2k-1} \rfloor$.

Proof If $|Y| \geq \lfloor \frac{kn}{2k-1} \rfloor$, then by the condition of Theorem 2.1.9 we have $N_G(Y) = V(G)$. That contradicts $N_G(Y) \neq V(G)$. This completes the proof of Claim 1. □

In terms of Claim 1 and the condition of Theorem 2.1.9, we obtain

$$n - m \geq |N_G(Y)| \geq \frac{2k-1}{k}|Y| = \frac{2k-1}{k}(n - |S|).$$

This inequality implies

$$|S| \geq \frac{(k-1)n + km}{2k-1}. \tag{2.1.28}$$

From (2.1.19), (2.1.28), $m \geq 1$ and the fact that $|S| + |T| \leq n$, we have

$$\begin{aligned}
-1 &\geq \delta_G(S,T) = k|S| + d_{G-S}(T) - k|T| \\
&\geq k|S| + |T| - m - k|T| \\
&= k|S| - (k-1)|T| - m \\
&\geq k|S| - (k-1)(n - |S|) - m \\
&= (2k-1)|S| - (k-1)n - m \\
&\geq (2k-1) \cdot \frac{(k-1)n + km}{2k-1} - (k-1)n - m \\
&= (k-1)m \geq k - 1 \geq 0.
\end{aligned}$$

This is a contradiction.

From all the cases above, we deduced the contradictions. Hence, G has a fractional k-factor. This completes the proof of Theorem 2.1.9. $\square$

Remark 2.1.2 Let us show that the condition in Theorem 2.1.9 cannot be replaced by the condition that $N_G(X) = V(G)$ or

$$|N_G(X)| \geq \frac{2k-1}{k}|X|$$

for all $X \subseteq V(G)$. Let k be an odd integer with $k \geq 2$. Let m be any odd positive integer. We construct a graph G of order n as follows. Let $V(G) = S \cup T$ (disjoint union), $|S| = (k-1)m$ and $|T| = km + 1$, and put $T = \{t_1, t_2, \cdots, t_{2l}\}$, where $2l = km + 1$. For each $s \in S$, define $N_G(s) = V(G) \setminus \{s\}$, and for any $t \in T$, define $N_G(t) = S \cup \{t'\}$, where $\{t, t'\} = \{t_{2i-1}, t_{2i}\}$ for some i, $1 \leq i \leq l$. Obviously, $n = (2k-1)m + 1$. We first show that the condition that $N_G(X) = V(G)$ or

$$|N_G(X)| \geq \frac{2k-1}{k}|X|$$

2.1 Fractional k-Factors

for all $X \subseteq V(G)$ holds. Let any $X \subseteq V(G)$. It is obvious that if $|X \cap S| \geq 2$, or $|X \cap S| = 1$ and $|X \cap T| \geq 1$, then

$$N_G(X) = V(G).$$

Of course, if $|X| = 1$ and $X \subseteq S$, then

$$|N_G(X)| = |V(G)| - 1 = n - 1 > \frac{n-1}{km} = \frac{(2k-1)m}{km}$$
$$= \frac{2k-1}{k} = \frac{2k-1}{k}|X|.$$

Hence, we may assume $X \subseteq T$. Since

$$|N_G(X)| = |S| + |X| = (k-1)m + |X|, \quad |N_G(X)| \geq \frac{2k-1}{k}|X|$$

holds if and only if

$$(k-1)m + |X| \geq \frac{2k-1}{k}|X|.$$

This inequality is equivalent to $|X| \leq km$. Thus if $X \neq T$ and $X \subset T$, then

$$|N_G(X)| \geq \frac{2k-1}{k}|X|$$

holds for all $X \subseteq V(G)$. If $X = T$, then

$$N_G(X) = V(G).$$

Consequently, $N_G(X) = V(G)$ or $|N_G(X)| \geq \frac{2k-1}{k}|X|$ for all $X \subseteq V(G)$ follows. In the following, we show that G has no fractional k-factor. For above S and T, obviously, $d_{G-S}(t) = 1$ for each $t \in T$. Thus, we obtain

$$\delta_G(S, T) = k|S| + d_{G-S}(T) - k|T|$$
$$= k|S| + |T| - k|T|$$
$$= k|S| - (k-1)|T|$$
$$= k(k-1)m - (k-1)(km+1)$$
$$= -(k-1) \leq -1.$$

In terms of Theorem 2.1.1, G has no fractional k-factor. In the above sense, the condition in Theorem 2.1.9 is best possible.

Katerinis and Woodall[8] gave a binding number condition for a graph to have a k-factor.

Theorem 2.1.10 [8] Let k be an integer such that $k \geq 2$, and let G be a graph of order n such that $n \geq 4k - 6$, kn is even, and

$$\text{bind}(G) > \frac{(2k-1)(n-1)}{k(n-2)+3}.$$

Then G has a k-factor.

D. R. Woodall[21] gave the following result.

Theorem 2.1.11 [21] Let G be a graph of order n with $\text{bind}(G) > c$. Then

$$\delta(G) > n - \frac{n-1}{c}.$$

Next we present a sufficient condition for graphs to have fractional k-factors by using the binding number. The result was verified by Zhou, Xu and Duan[45].

Theorem 2.1.12 [45] Let $k \geq 2$ be an integer, and let G be a graph of order n such that $n \geq 4k - 6$. Then

(1) If kn is even, and $\text{bind}(G) > \frac{(2k-1)(n-1)}{k(n-2)+3}$, then G has a fractional k-factor;

(2) If kn is odd, and $\text{bind}(G) > \frac{(2k-1)(n-1)}{k(n-2)+2}$, then G has a fractional k-factor.

Proof If kn is even, and $\text{bind}(G) > \frac{(2k-1)(n-1)}{k(n-2)+3}$. By Theorem 2.1.10, G has a k-factor. We have known that a k-factor is a special fractional k-factor. Thus, G has a fractional k-factor. In the following, we prove (2).

Suppose that G does not have a fractional k-factor. Then, according to Theorem 2.1.1, there exists some $S \subseteq V(G)$ such that

$$\delta_G(S, T) = k|S| - k|T| + d_{G-S}(T) \leq -1, \qquad (2.1.29)$$

where

2.1 Fractional k-Factors

$$T = \{x : x \in V(G) \setminus S, \ d_{G-S}(x) \leq k\}.$$

We choose such subsets S and T so that $|T|$ is as small as possible.

Claim 1 $|T| \geq k+1$.

Proof According to Theorem 2.1.11, we have

$$|S| + d_{G-S}(x) \geq d_G(x) \geq \delta(G) > n - \frac{n-1}{\text{bind}(G)}$$

$$\geq \frac{n(k-1) + 2k - 3}{2k - 1}$$

$$\geq \frac{(4k-6)(k-1) + 2k - 3}{2k - 1}$$

$$\geq 2k - 3 \geq k \quad \text{(since } k \geq 2 \text{ is odd)}.$$

If $|T| \leq k$, then by (2.1.29) we obtain

$$-1 \geq \delta_G(S, T) = k|S| + d_{G-S}(T) - k|T|$$

$$\geq |T||S| + d_{G-S}(T) - k|T|$$

$$= \sum_{x \in T}(|S| + d_{G-S}(x) - k) \geq 0,$$

which is a contradiction. □

Claim 2 $d_{G-S}(x) \leq k - 1$ for all $x \in T$.

Proof If $d_{G-S}(x) \geq k$ for some $x \in T$, then the subsets S and $T \setminus \{x\}$ satisfy (2.1.29). This contradicts the choice of S and T. □

Define

$$h = \min\{d_{G-S}(x) | x \in T\}.$$

Then by Claim 2, we have $0 \leq h \leq k - 1$.

Choose $x_1 \in T$ such that $d_{G-S}(x_1) = h$. The proof splits into two cases.

Case 1 $1 \leq h \leq k - 1$.

Let $Y = (V(G) \setminus S) \setminus N_{G-S}(x_1)$. Then $x_1 \in Y \setminus N_G(Y)$, so $Y \neq \emptyset$ and $N_G(Y) \neq V(G)$, and $|N_G(Y)| \geq \text{bind}(G)|Y|$. Thus, we obtain

$$n - 1 \geq |N_G(Y)| \geq \text{bind}(G)|Y| = \text{bind}(G)(n - |S| - h),$$

i.e.,

$$|S| \geq n - h - \frac{n-1}{\text{bind}(G)} > n - h - \frac{k(n-2)+2}{2k-1}. \quad (2.1.30)$$

Subcase 1.1 $3 \leq h \leq k-1$.

By (2.1.29) and (2.1.30), and $|T| \leq n - |S|$, we have

$$-1 \geq \delta_G(S,T) = k|S| + d_{G-S}(T) - k|T|$$
$$\geq k|S| - k|T| + h|T| = k|S| - (k-h)|T|$$
$$\geq k|S| - (k-h)(n-|S|)$$
$$= (2k-h)|S| - kn + hn$$
$$> (2k-h)\left(n - h - \frac{k(n-2)+2}{2k-1}\right) - kn + hn$$
$$= kn - (2k-h)h - (2k-h)\frac{k(n-2)+2}{2k-1}.$$

Let

$$f(h) = kn - (2k-h)h - (2k-h)\frac{k(n-2)+2}{2k-1}.$$

Then

$$f'(h) = -2k + 2h + \frac{k(n-2)+2}{2k-1}.$$

Since $3 \leq h \leq k-1$, we have

$$f'(h) \geq -2k + 6 + \frac{k(n-2)+2}{2k-1}$$
$$= \frac{-4k^2 + 2k + 12k - 6 + kn - 2k + 2}{2k-1}$$
$$= \frac{kn - 4k^2 + 12k - 4}{2k-1}$$
$$\geq \frac{k(4k-6) - 4k^2 + 12k - 4}{2k-1}$$
$$= \frac{6k-4}{2k-1} > 0.$$

Thus, we get $f(h) \geq f(3)$, i.e.,

$$-1 > f(h) \geq f(3) = kn - 3(2k-3) - (2k-3)\frac{k(n-2)+2}{2k-1}.$$

2.1 Fractional k-Factors

$$= \frac{2k^2n - kn - (6k-9)(2k-1) - (2k-3)(k(n-2)+2)}{2k-1}$$

$$= \frac{2kn - 8k^2 + 14k - 3}{2k-1} \geq \frac{2k(4k-6) - 8k^2 + 14k - 3}{2k-1}$$

$$= \frac{2k-3}{2k-1} > 0,$$

which is a contradiction.

Subcase 1.2 $h = 2$.

Claim 3 $(k-2)(2k-1)|T| \leq k[(n-2)(2k-1) - (k(n-2)+2)] + (2k-1)$, that is,

$$|T| \leq \frac{k}{k-2}\left(n - 2 - \frac{k(n-2)+2}{2k-1}\right) + \frac{1}{k-2}.$$

Proof If

$$(k-2)(2k-1)|T| \geq k[(n-2)(2k-1) - (k(n-2)+2)] + (2k-1) + 1,$$

that is,

$$|T| \geq \frac{k}{k-2}\left(n - 2 - \frac{k(n-2)+2}{2k-1}\right) + \frac{1}{k-2} + \frac{1}{(2k-1)(k-2)}.$$

Then, by (2.1.30), we obtain

$$|S| + |T| > n - 2 - \frac{k(n-2)+2}{2k-1} + \frac{k}{k-2}\left(n - 2 - \frac{k(n-2)+2}{2k-1}\right)$$

$$+ \frac{1}{k-2} + \frac{1}{(2k-1)(k-2)}$$

$$= \frac{2k-2}{k-2}\left(n - 2 - \frac{k(n-2)+2}{2k-1}\right) + \frac{1}{k-2} + \frac{1}{(2k-1)(k-2)}$$

$$= n + \frac{kn - 4k^2 + 6k - 1}{(2k-1)(k-2)} + \frac{1}{(2k-1)(k-2)}$$

$$\geq n + \frac{k(4k-6) - 4k^2 + 6k - 1}{(2k-1)(k-2)} + \frac{1}{(2k-1)(k-2)}$$

$$= n - \frac{1}{(2k-1)(k-2)} + \frac{1}{(2k-1)(k-2)} = n.$$

This contradicts $|S| + |T| \leq n$. $\square$

By combining Claim 3 with (2.1.29) and (2.1.30), we obtain

$$-1 \geq \delta_G(S,T) = k|S| + d_{G-S}(T) - k|T|$$
$$\geq k|S| - k|T| + 2|T| = k|S| - (k-2)|T|$$
$$> k\left(n - 2 - \frac{k(n-2)+2}{2k-1}\right)$$
$$- (k-2)\left(\frac{k}{k-2}\left(n - 2 - \frac{k(n-2)+2}{2k-1}\right) + \frac{1}{k-2}\right)$$
$$= -1,$$

a contradiction.

Subcase 1.3 $h = 1$.

Subcase 1.3.1 $|T| \leq \frac{k}{k-1}\left(n - 1 - \frac{k(n-2)+2}{2k-1}\right) + \frac{1}{k-1}$.

By combining this with (2.1.29) and (2.1.30), we have that

$$-1 \geq \delta_G(S,T) = k|S| + d_{G-S}(T) - k|T|$$
$$\geq k|S| - k|T| + |T| = k|S| - (k-1)|T|$$
$$> k\left(n - 1 - \frac{k(n-2)+2}{2k-1}\right)$$
$$- (k-1)\left(\frac{k}{k-1}\left(n - 1 - \frac{k(n-2)+2}{2k-1}\right) + \frac{1}{k-1}\right)$$
$$= -1,$$

a contradiction.

Subcase 1.3.2 $|T| > \frac{k}{k-1}\left(n - 1 - \frac{k(n-2)+2}{2k-1}\right) + \frac{1}{k-1}$.

In view of (2.1.30), we obtain

$$|S| + |T| > n - 1 - \frac{k(n-2)+2}{2k-1}$$
$$+ \frac{k}{k-1}\left(n - 1 - \frac{k(n-2)+2}{2k-1}\right) + \frac{1}{k-1}$$
$$> \frac{2k-1}{k-1}\left(n - 1 - \frac{k(n-2)+2}{2k-1}\right) + \frac{1}{k-1}$$

2.1 Fractional k-Factors

$$= \frac{kn-n-1}{k-1} + \frac{1}{k-1} = n.$$

This contradicts $|S|+|T| \leq n$.

Case 2 $h = 0$.

Let m be the number of vertices x in T such that $d_{G-S}(x) = 0$, and let $Y = V(G) \setminus S$. Then $N_G(Y) \neq V(G)$ since $h = 0$, and $Y \neq \emptyset$ by Claim 1, and so $|N_G(Y)| \geq \text{bind}(G)|Y|$. Thus

$$n - m \geq |N_G(Y)| \geq \text{bind}(G)|Y| = \text{bind}(G)(n-|S|).$$

So

$$|S| \geq n - \frac{n-m}{\text{bind}(G)}. \tag{2.1.31}$$

In view of (2.1.29) and (2.1.31), and $|T| \leq n - |S|$, we get that

$$-1 \geq \delta_G(S,T) = k|S| + d_{G-S}(T) - k|T|$$
$$\geq k|S| - k|T| + |T| - m$$
$$\geq k|S| - (k-1)(n-|S|) - m$$
$$= (2k-1)|S| - kn + n - m$$
$$\geq (2k-1)\left(n - \frac{n-m}{\text{bind}(G)}\right) - kn + n - m$$
$$= 2kn - n - \frac{n(2k-1)}{\text{bind}(G)} + \frac{m(2k-1)}{\text{bind}(G)} - kn + n - m$$
$$= kn - \frac{n(2k-1)}{\text{bind}(G)} + \frac{m(2k-1)}{\text{bind}(G)} - m$$
$$\geq kn - \frac{n(2k-1)}{\text{bind}(G)} + \frac{2k-1}{\text{bind}(G)} - 1$$
$$= kn - \frac{(n-1)(2k-1)}{\text{bind}(G)} - 1$$
$$> kn - (k(n-2) + 2) - 1 = 2k - 3 > 0.$$

This is a contradiction.

From all the cases above, we deduced the contradiction. Hence, G has a fractional k-factor. Completing the proof of Theorem 2.1.12. □

Remark 2.1.3 Let kn be even. Then, let us show that the condition

$$\text{bind}(G) > \frac{(2k-1)(n-1)}{k(n-2)+3}$$

in Theorem 2.1.12 can not be replaced by

$$\text{bind}(G) \geq \frac{(2k-1)(n-1)}{k(n-2)+3}.$$

Let r be a positive integer and let $l = rk - 1$ and $m = 2l - 2r$, so that $km = 2l(k-1) - 2$ and $n = m + 2l = (4k-2)r - 4$. Let $H = K_m \bigvee lK_2$. Let $X = V(lK_2)$. Then for any $x \in X$, $|N_H(X \setminus x)| = n - 1$. By the definition of $\text{bind}(H)$,

$$\text{bind}(H) = \frac{|N_H(X \setminus x)|}{|X \setminus x|} = \frac{n-1}{2l-1} = \frac{n-1}{2rk-3}$$

$$= \frac{(2k-1)(n-1)}{k(n-2)+3}.$$

Let $S = V(K_m) \subseteq V(H)$, $T = V(lK_2) \subseteq V(H)$. Then $|S| = m$, $|T| = 2l$. Thus, we get

$$\delta_H(S,T) = k|S| - k|T| + d_{H-S}(T)$$
$$= k|S| - k|T| + |T| = k|S| - (k-1)|T|$$
$$= km - 2(k-1)l = -2 < 0.$$

By Theorem 2.1.1, there are not any fractional k-factors in H. In the above sense, the result in Theorem 2.1.12 is best possible.

Let kn be odd. Then, let us show that the condition

$$\text{bind}(G) > \frac{(2k-1)(n-1)}{k(n-2)+2}$$

in Theorem 2.1.12 can not be replaced by

$$\text{bind}(G) \geq \frac{(2k-1)(n-1)}{k(n-2)+2}.$$

Let $r \geq 1$, $k \geq 3$ be two odd positive integer and let $l = \frac{5kr-1}{2}$ and $m = 5kr - 5r - 1$, so that

$$n = m + 2l = (10k - 5)r - 2.$$

2.1 Fractional k-Factors

Clearly, n is odd. Let $H = K_m \vee lK_2$. Let $X = V(lK_2)$. Then for any $x \in X$, $|N_H(X \setminus x)| = n - 1$. By the definition of bind(H),

$$\text{bind}(H) = \frac{|N_H(X \setminus x)|}{|X \setminus x|} = \frac{n-1}{2l-1} = \frac{n-1}{5kr-2}$$

$$= \frac{(2k-1)(n-1)}{k(n-2)+2}.$$

Let $S = V(K_m) \subseteq V(H)$, $T = V(lK_2) \subseteq V(H)$. Then $|S| = m$, $|T| = 2l$. Thus, we get

$$\delta_H(S,T) = k|S| - k|T| + d_{H-S}(T) = k|S| - k|T| + |T|$$
$$= k|S| - (k-1)|T| = km - 2(k-1)l$$
$$= k(5kr - 5r - 1) - (k-1)(5kr - 1)$$
$$= -1 < 0.$$

By Theorem 2.1.1, there are not any fractional k-factors in H. In the above sense, the result in Theorem 2.1.12 is best possible.

Niessen[16] proved a result on the existence of k-factors in graphs.

Theorem 2.1.13 [16] Let $k \geq 2$ be an integer and G a graph with n vertices. Assume that if k is odd, then n is even and G is connected. Let G satisfy $n > 4k + 1 - 4\sqrt{k+2}$,

$$\delta(G) \geq \frac{(k-1)(n+2)}{2k-1} \quad \text{and}$$

$$\delta(G) > \frac{1}{2k-2}((k-2)n + 2\alpha(G) - 2).$$

Then G has a k-factor.

In the following, we show two sufficient conditions for the existence of fractional k-factors in graphs by using the independence number and minimum degree. The results were obtained by Zhou[36].

Theorem 2.1.14 [36] Let $k \geq 2$ be an even integer and G a graph of order n with $n > 4k + 1 - 4\sqrt{k+2}$. If

$$\delta(G) \geq \frac{(k-1)(n+2)}{2k-1} \quad \text{and}$$

$$\delta(G) > \frac{1}{2k-2}((k-2)n + 2\alpha(G) - 2),$$

then G has a fractional k-factor.

Proof According to Theorem 2.1.13, G has a k-factor. In view of the definitions of a k-factor and a fractional k-factor, we know that a k-factor is a special fractional k-factor. Hence, G has a fractional k-factor. The proof of Theorem 2.1.14 is complete. □

Remark 2.1.4 Let us show that the condition

$$\delta(G) \geq \frac{(k-1)(n+2)}{2k-1}$$

in Theorem 2.1.14 can not be replaced by

$$\delta(G) \geq \frac{(k-1)(n+2) - 1}{2k-1}.$$

Let $G = K_{2k-4} \vee (k-1)K_2$, and we denote by n the order of graph G. Then $n = 4k - 6 > 4k + 1 - 4\sqrt{k+2}$ and

$$\delta(G) = 2k - 3 = \frac{(k-1)(n+2) - 1}{2k-1}.$$

On the other hand, $\alpha(G) = k - 1$ and

$$\delta(G) = 2k - 3 = \frac{(2k-3)(2k-2)}{2k-2} = \frac{4k^2 - 10k + 6}{2k-2}$$

$$> \frac{4k^2 - 12k + 8}{2k-2} = \frac{(k-2)(4k-6) + 2(k-1) - 2}{2k-2}$$

$$= \frac{(k-2)n + 2\alpha(G) - 2}{2k-2}.$$

Let

$$S = V(K_{2k-4}) \subseteq V(G), \quad T = V((k-1)K_2) \subseteq V(G).$$

Then $|S| = 2k - 4$, $|T| = 2k - 2$, and

$$d_{G-S}(T) = 2k - 2.$$

Thus, we obtain

$$\delta_G(S,T) = k|S| + d_{G-S}(T) - k|T|$$
$$= k(2k-4) + 2k - 2 - k(2k-2)$$
$$= -2 < 0.$$

Then by Theorem 2.1.1, there are no fractional k-factors in G. In the above sense, the condition
$$\delta(G) \geq \frac{(k-1)(n+2)}{2k-1}$$
in Theorem 2.1.14 is best possible.

Let us show that the condition
$$\delta(G) > \frac{1}{2k-2}((k-2)n + 2\alpha(G) - 2)$$
in Theorem 2.1.14 can not be replaced by
$$\delta(G) \geq \frac{1}{2k-2}((k-2)n + 2\alpha(G) - 2).$$

Since $k \geq 2$ is even, we have that $\frac{k-2}{2}$ is a nonnegative integer. Let
$$G = K_{3k-3} \vee \left(2kK_1 \bigcup \frac{k-2}{2} K_2\right),$$
and we use n to denote the order of graph G. Obviously,
$$n = 6k - 5 > 4k + 1 - 4\sqrt{k+2}$$
and $\alpha(G) = 2k + \frac{k-2}{2}$. Thus,
$$\delta(G) = 3k - 3 = \frac{(3k-3)(2k-2)}{2k-2} = \frac{6k^2 - 12k + 6}{2k-2}$$
$$= \frac{(k-2)(6k-5) + 2(2k + \frac{k-2}{2}) - 2}{2k-2}$$
$$= \frac{(k-2)n + 2\alpha(G) - 2}{2k-2}$$
and
$$\delta(G) = 3k - 3 = \frac{(3k-3)(2k-1)}{2k-1} = \frac{(k-1)(6k-3)}{2k-1}$$
$$= \frac{(k-1)(n+2)}{2k-1}.$$

Let
$$S = V(K_{3k-3}) \subseteq V(G), \quad T = V\left(2kK_1 \bigcup \frac{k-2}{2}K_2\right) \subseteq V(G).$$
Then $|S| = 3k - 3$, $|T| = 3k - 2$, and $d_{G-S}(T) = k - 2$. Thus, we have
$$\begin{aligned}\delta_G(S,T) &= k|S| + d_{G-S}(T) - k|T| \\ &= k(3k-3) + k - 2 - k(3k-2) \\ &= -2 < 0.\end{aligned}$$

Then by Theorem 2.1.1, there are no fractional k-factors in G. In the above sense, the condition
$$\delta(G) > \frac{1}{2k-2}((k-2)n + 2\alpha(G) - 2)$$
in Theorem 2.1.14 is best possible.

Theorem 2.1.15 [36] *Let $k \geq 3$ be an odd integer and G a graph of order n with $n \geq 4k - 5$. If*
$$\delta(G) > \frac{(k-1)(n+2)}{2k-1} \quad \text{and}$$
$$\delta(G) > \frac{1}{2k-2}((k-2)n + 2\alpha(G) - 1),$$
then G has a fractional k-factor.

Proof Suppose that a graph G satisfies the conditions of Theorem 2.1.15, but it does not have a fractional k-factor. Then by Theorem 2.1.1, there exists some $S \subseteq V(G)$ such that
$$\delta_G(S,T) = k|S| + d_{G-S}(T) - k|T| \leq -1, \tag{2.1.32}$$
where $T = \{x : x \in V(G) \setminus S, d_{G-S}(x) \leq k\}$. We choose disjoint subsets S and T such that (2.1.32) holds, $d_{G-S}(x) \leq k$ for all $x \in T$, and $|T|$ is as small as possible.

If $T = \emptyset$, then by (2.1.32),
$$-1 \geq \delta_G(S,T) = k|S| \geq 0,$$
a contradiction. Hence, $T \neq \emptyset$.

2.1 Fractional k-Factors

Claim 1 $d_{G-S}(x) \leq k - 1$ for each $x \in T$.

Proof If $d_{G-S}(x) \geq k$ for some $x \in T$, then the subsets S and $T \setminus \{x\}$ satisfy (1). This contradicts the choice of S and T. □

Since $T \neq \emptyset$, we may choose a vertex $x_1 \in T$ with

$$h = \min\{d_{G-S}(x) : x \in T\} = d_{G-S}(x_1).$$

Clearly, we have

$$\delta(G) \leq d_G(x_1) \leq d_{G-S}(x_1) + |S| = h + |S|. \tag{2.1.33}$$

In view of Claim 1, we get

$$0 \leq h \leq k - 1.$$

We shall consider two cases by the value of h and derive contradictions.

Case 1 $1 \leq h \leq k - 1$.

In view of (2.1.32), (2.1.33) and $|S| + |T| \leq n$, we obtain

$$-1 \geq \delta_G(S, T) = k|S| + d_{G-S}(T) - k|T|$$
$$\geq k|S| - (k - h)|T| \geq k|S| - (k - h)(n - |S|)$$
$$= (2k - h)|S| - (k - h)n$$
$$\geq (2k - h)(\delta(G) - h) - (k - h)n$$
$$> (2k - h)\left(\frac{(k-1)(n+2)}{2k-1} - h\right) - (k - h)n,$$

that is,

$$-1 > (2k - h)\left(\frac{(k-1)(n+2)}{2k-1} - h\right) - (k - h)n. \tag{2.1.34}$$

Subcase 1.1 $h = 1$.

In view of (2.1.34), we have

$$-1 > (2k - 1)\left(\frac{(k-1)(n+2)}{2k-1} - 1\right) - (k - 1)n = -1.$$

This is a contradiction.

Subcase 1.2 $2 \leq h \leq k - 1$.

Let

$$f(h) = (2k - h)\left(\frac{(k-1)(n+2)}{2k-1} - h\right) - (k - h)n.$$

39

Then by $n \geq 4k - 5$, we obtain

$$f'(h) = 2h - \frac{(k-1)(n+2)}{2k-1} - 2k + n$$

$$\geq 4 - \frac{(k-1)(n+2)}{2k-1} - 2k + n$$

$$= \frac{kn - 2(k-1)}{2k-1} - 2k + 4$$

$$\geq \frac{k(4k-5) - 2(k-1)}{2k-1} - 2k + 4$$

$$= \frac{3k-2}{2k-1} > 0.$$

According to $2 \leq h \leq k - 1$ and $f'(h) > 0$, we have

$$f(h) \geq f(2). \tag{2.1.35}$$

In view of (2.1.34), (2.1.35) and $n \geq 4k - 5$, we obtain

$$-1 > f(2) = (2k-2)\left(\frac{(k-1)(n+2)}{2k-1} - 2\right) - (k-2)n$$

$$= \frac{kn - 2k(2k-2)}{2k-1} \geq \frac{k(4k-5) - 2k(2k-2)}{2k-1}$$

$$= -\frac{k}{2k-1} > -1,$$

which is a contradiction.

Case 2 $h = 0$.

Let $X = \{x \in T : d_{G-S}(x) = 0\}$, $Y = \{x \in T : d_{G-S}(x) = 1\}$, $Y_1 = \{x \in Y : N_{G-S}(x) \subseteq T\}$ and $Y_2 = Y - Y_1$. Then the graph induced by Y_1 in $G - S$ has maximum degree at most 1. Let Z be a maximum independent set of this graph. Clearly,

$$|Z| \geq \frac{1}{2}|Y_1|.$$

In view of our definitions, $X \cup Z \cup Y_2$ is an independent set of G. Thus, we obtain

$$\alpha(G) \geq |X| + |Z| + |Y_2| \geq |X| + \frac{1}{2}|Y_1| + \frac{1}{2}|Y_2| = |X| + \frac{1}{2}|Y|.$$

$$\tag{2.1.36}$$

By $|S|+|T| \leq n$, (2.1.32), (2.1.33) and (2.1.36), we have

$$
\begin{aligned}
-1 \geq \delta_G(S,T) &= k|S| + d_{G-S}(T) - k|T| \\
&= k|S| + d_{G-S}(T \setminus (X \cup Y)) - k|T| + |Y| \\
&\geq k|S| + 2|T - (X \cup Y)| - k|T| + |Y| \\
&= k|S| + 2|T| - k|T| - 2|X| - |Y| \\
&= k|S| - (k-2)|T| - 2\left(|X| + \frac{1}{2}|Y|\right) \\
&\geq k|S| - (k-2)(n - |S|) - 2\left(|X| + \frac{1}{2}|Y|\right) \\
&= (2k-2)|S| - (k-2)n - 2\left(|X| + \frac{1}{2}|Y|\right) \\
&\geq (2k-2)\delta(G) - (k-2)n - 2\alpha(G) \\
&> (k-2)n + 2\alpha(G) - 1 - (k-2)n - 2\alpha(G) \\
&= -1,
\end{aligned}
$$

this is a contradiction.

From the contradictions we deduce that G has a fractional k-factor. This completes the proof of Theorem 2.1.15. □

Remark 2.1.5 Let us show that the condition

$$\delta(G) > \frac{(k-1)(n+2)}{2k-1}$$

in Theorem 2.1.15 can not be replaced by

$$\delta(G) \geq \frac{(k-1)(n+2)}{2k-1}.$$

Since $k \geq 3$ is odd, we have that $\frac{3k-1}{2}$ is a positive integer. Let

$$G = K_{3k-4} \vee \frac{3k-1}{2} K_2,$$

and we use n to denote the order of graph G. Then $n = 6k - 5 > 4k - 5$ and

$$\delta(G) = 3k - 3 = \frac{(k-1)(n+2)}{2k-1}.$$

Clearly, $\alpha(G) = \frac{3k-1}{2}$ and

Chapter 2 Fractional Factors

$$\delta(G) = 3k - 3 = \frac{(3k-3)(2k-2)}{2k-2} = \frac{6k^2 - 12k + 6}{2k-2}$$

$$> \frac{6k^2 - 14k + 8}{2k-2} = \frac{(k-2)(6k-5) + (3k-1) - 1}{2k-2}$$

$$= \frac{(k-2)n + 2\alpha(G) - 1}{2k-2}.$$

Let
$$S = V(K_{3k-4}) \subseteq V(G), \quad T = V\left(\frac{3k-1}{2}K_2\right) \subseteq V(G).$$

Then $|S| = 3k-4$, $|T| = 3k-1$, and $d_{G-S}(T) = 3k-1$. Thus, we obtain

$$\delta_G(S,T) = k|S| + d_{G-S}(T) - k|T|$$
$$= k(3k-4) + 3k - 1 - k(3k-1)$$
$$= -1 < 0.$$

Then by Theorem 2.1.1, there are no fractional k-factors in G. In the above sense, the condition

$$\delta(G) > \frac{(k-1)(n+2)}{2k-1}$$

in Theorem 2.1.15 is best possible.

Let us show that the condition

$$\delta(G) > \frac{1}{2k-2}((k-2)n + 2\alpha(G) - 1)$$

in Theorem 2.1.15 can not be replaced by

$$\delta(G) \geq \frac{1}{2k-2}((k-2)n + 2\alpha(G) - 1).$$

Since $k \geq 3$ is odd, we have that $\frac{k-1}{2}$ is a positive integer. Let

$$G = K_{3k-2} \bigvee \left(2kK_1 \bigcup \frac{k-1}{2}K_2\right),$$

and we use n to denote the order of graph G. Obviously, $n = 6k - 3 > 4k - 5$ and $\alpha(G) = 2k + \frac{k-1}{2}$. Thus,

$$\delta(G) = 3k - 2 = \frac{(3k-2)(2k-2)}{2k-2} = \frac{6k^2 - 10k + 4}{2k-2}$$

2.1 Fractional k-Factors

$$= \frac{(k-2)(6k-3) + 2\left(2k + \frac{k-1}{2}\right) - 1}{2k-2}$$

$$= \frac{(k-2)n + 2\alpha(G) - 1}{2k-2}$$

and

$$\delta(G) = 3k - 2 = \frac{(3k-2)(2k-1)}{2k-1} = \frac{6k^2 - 7k + 2}{2k-1}$$

$$> \frac{6k^2 - 7k + 1}{2k-1} = \frac{(k-1)(6k-1)}{2k-1}$$

$$= \frac{(k-1)(n+2)}{2k-1}.$$

Let $S = V(K_{3k-2}) \subseteq V(G)$, $T = V(2kK_1 \cup \frac{k-1}{2}K_2) \subseteq V(G)$. Then $|S| = 3k-2$, $|T| = 3k-1$, and $d_{G-S}(T) = k-1$. Thus, we have

$$\delta_G(S,T) = k|S| + d_{G-S}(T) - k|T|$$
$$= k(3k-2) + k - 1 - k(3k-1)$$
$$= -1 < 0.$$

Then by Theorem 2.1.1, there are no fractional k-factors in G. In the above sense, the condition

$$\delta(G) > \frac{1}{2k-2}((k-2)n + 2\alpha(G) - 1)$$

in Theorem 2.1.15 is best possible.

Finally, we give two toughness condition for graphs to have fractional factors.

Lemma 2.1.4 [6] Suppose that G is not a complete graph. Then $\tau(G) \le \kappa(G)$. In particular, G is connected if and only if $\tau(G) > 0$.

Lemma 2.1.5 [3] For a graph G, we have

$$\kappa(G) \le \lambda(G) \le \delta(G).$$

Theorem 2.1.16 [38] Let G be a graph with $\tau(G) > 1$. Then G contains a fractional 1-factor.

Theorem 2.1.17 [38]   *Let G be a graph with $\tau(G) > 2$. Then G contains a fractional 2-factor.*

Proof   The theorem is true for a complete graph. In the following we assume that G is not a complete graph. Then by Theorem 2.1.2, we only need prove
$$2p_0(G-S) + p_1(G-S) \leq 2|S|$$
for any $S \subseteq V(G)$.

In terms of Lemma 2.1.4 and Lemma 2.1.5, we have
$$\delta(G) \geq \lambda(G) \geq \kappa(G) \geq \tau(G) > 2.$$
If $S = \emptyset$, then we obtain $p_0(G-S) = p_1(G-S) = 0$, and so
$$2p_0(G-S) + p_1(G-S) = 0 \leq 2|S|.$$

In the following we consider $S \neq \emptyset$. Assume that G has no fractional 2-factor. Then by Theorem 2.1.2, there exists some $\emptyset \neq S_0 \subseteq V(G)$ such that
$$2p_0(G-S_0) + p_1(G-S_0) > 2|S_0|. \tag{2.1.37}$$

We use $i(G-S_0)$ to denote the number of isolated vertices of $G-S_0$, i.e., $i(G-S_0) = p_0(G-S_0)$. We write
$$T_1 = \{x : x \in V(G) \setminus S_0, d_{G-S_0}(x) = 1\},$$
$$N_{G-S_0}(T_1) = \{y : xy \in E_{G-S_0}(G), x \in T_1, y \notin T_1\},$$
$$E(T_1) = \{e = uv : u \neq v, u, v \in T_1\},$$
$$E(T_1, N_{G-S_0}(T_1)) = \{e = uv : u \in T_1, v \notin T_1, v \in N_{G-S_0}(T_1)\}.$$

If $E(T_1) \neq \emptyset$, then we write $E(T_1) = \{x_1y_1, x_2y_2, \cdots, x_ky_k\}$ and $T_{1/2} = \{x_1, x_2, \cdots, x_k\}$. We shall consider four cases.

Case 1 $E(T_1) = \emptyset$ and $E(T_1, N_{G-S_0}(T_1)) = \emptyset$.

In this case, we have $p_1(G-S_0) = 0$. Combining this with (2.1.37), we obtain
$$\omega(G-S_0) \geq i(G-S_0) = p_0(G-S_0) \geq |S_0| + 1 \geq 2.$$
According to the condition of the theorem and the definition of $\tau(G)$, we obtain

$$2 < \tau(G) \leq \frac{|S_0|}{\omega(G - S_0) - 1} \leq 1,$$

which is a contradiction.

Case 2 $E(T_1) = \emptyset$ and $E(T_1, N_{G-S_0}(T_1)) \neq \emptyset$.

Set $V_1 = N_{G-S_0}(T_1)$. Then

$$0 < |V_1| \leq |T_1| = p_1(G - S_0).$$

It follows from (2.1.37) that

$$\omega(G - S_0 \cup V_1) = \omega(G - S_0 - V_1) \geq i(G - S_0 - V_1)$$
$$\geq i(G - S_0) + p_1(G - S_0)$$
$$> i(G - S_0) + \frac{p_1(G - S_0)}{2}$$
$$> |S_0| \geq 1. \qquad (2.1.38)$$

If $i(G - S_0) = 0$, then by (2.1.38) we have

$$\omega(G - S_0 - V_1) \geq p_1(G - S_0) > \frac{p_1(G - S_0)}{2} > |S_0| \geq 1,$$

and so

$$\omega(G - S_0 - V_1) \geq p_1(G - S_0) \geq 2|S_0| + 1 \geq 3.$$

According to the condition of the theorem and the definition of $\tau(G)$, we obtain

$$2 < \tau(G) \leq \frac{|S_0 \cup V_1|}{\omega(G - S_0 \cup V_1) - 1} = \frac{|S_0| + |V_1|}{\omega(G - S_0 - V_1) - 1}$$
$$\leq \frac{(\omega(G - S_0 - V_1) - 1)/2 + \omega(G - S_0 - V_1)}{\omega(G - S_0 - V_1) - 1}$$
$$= \frac{3}{2} + \frac{1}{\omega(G - S_0 - V_1) - 1} \leq \frac{3}{2} + \frac{1}{3 - 1}$$
$$= 2,$$

which is a contradiction.

If $i(G - S_0) \geq 1$, then by (2.1.38) we obtain

$$p_1(G - S_0) \leq \omega(G - S_0 - V_1) - i(G - S_0)$$
$$\leq \omega(G - S_0 - V_1) - 1.$$

In view of the condition of the theorem and the definition of $\tau(G)$, we have

$$2 < \tau(G) \leq \frac{|S_0 \cup V_1|}{\omega(G - S_0 \cup V_1) - 1} = \frac{|S_0| + |V_1|}{\omega(G - S_0 - V_1) - 1}$$

$$\leq \frac{|S_0| + p_1(G - S_0)}{\omega(G - S_0 - V_1) - 1}$$

$$\leq \frac{(\omega(G - S_0 - V_1) - 1) + (\omega(G - S_0 - V_1) - 1)}{\omega(G - S_0 - V_1) - 1}$$

$$= 2,$$

which is a contradiction.

Case 3 $E(T_1) \neq \emptyset$ and $E(T_1, N_{G-S_0}(T_1)) = \emptyset$.

We write $V_1 = T_{1/2}$. Then we have

$$|V_1| = |T_{1/2}| = \frac{p_1(G - S_0)}{2}.$$

In terms of (2.1.37), we obtain

$$\omega(G - S_0 \cup V_1) = \omega(G - S_0 - V_1) \geq i(G - S_0 - V_1)$$

$$\geq i(G - S_0) + \frac{p_1(G - S_0)}{2}$$

$$> |S_0| \geq 1.$$

According to the definition of $T_{1/2}$ and the choice of V_1, it is easy to see that

$$\omega(G - S_0) = \omega(G - S_0 \cup V_1) = \omega(G - S_0 - V_1) \geq |S_0| + 1 \geq 2.$$

By the condition of the theorem and the definition of $\tau(G)$, we obtain

$$2 < \tau(G) \leq \frac{|S_0|}{\omega(G - S_0) - 1} \leq 1,$$

which is a contradiction.

Case 4 $E(T_1) \neq \emptyset$ and $E(T_1, N_{G-S_0}(T_1)) \neq \emptyset$.

Subcase 4.1 $|E(T_1)| > |E(T_1, N_{G-S_0}(T_1))|$.

We choose $V_1 = N_{G-S_0}(T_1) \cup T'$ such that

$$|V_1| = \left\lfloor \frac{p_1(G - S_0)}{2} \right\rfloor,$$

where $T' \subseteq T_{1/2}$. It follows from (2.1.37) that

2.1 Fractional k-Factors

$$\omega(G - S_0 \cup V_1) = \omega(G - S_0 - V_1) \geq i(G - S_0 - V_1)$$
$$\geq i(G - S_0) + \frac{p_1(G - S_0)}{2}$$
$$> |S_0| \geq 1.$$

In terms of $E(T_1) \neq \emptyset$, $|E(T_1)| > |E(T_1, N_{G-S_0}(T_1))|$ and the choice of V_1, it is easy to see that there exists at least one vertex x with $x \in T_{1/2}$ and $x \in V_1$. Combining this with the definition of $T_{1/2}$, we obtain

$$\omega(G - S_0 \cup (V_1 \setminus x)) = \omega(G - S_0 - (V_1 \setminus x)) = \omega(G - S_0 - V_1)$$
$$\geq |S_0| + 1 \geq 2.$$

According to the condition of the theorem and the definition of $\tau(G)$, we have

$$2 < \tau(G) \leq \frac{|S_0 \cup (V_1 \setminus x)|}{\omega(G - S_0 \cup (V_1 \setminus x)) - 1} = \frac{|S_0| + |V_1| - 1}{\omega(G - S_0 - V_1) - 1}$$
$$\leq \frac{\omega(G - S_0 - V_1) - 1 + p_1(G - S_0)/2 - 1}{\omega(G - S_0 - V_1) - 1}$$
$$\leq \frac{\omega(G - S_0 - V_1) - 1 + \omega(G - S_0 - V_1) - 1}{\omega(G - S_0 - V_1) - 1} = 2,$$

which is a contradiction.

Subcase 4.2 $|E(T_1)| \leq |E(T_1, N_{G-S_0}(T_1))|$.

We choose $V_1 = T_{1/2} \cup T'$ satisfying $|V_1| = \lfloor p_1(G - S_0)/2 \rfloor$, where $T' \subseteq N_{G-S_0}(T_1)$. By (2.1.37) we have

$$\omega(G - S_0 \cup V_1) = \omega(G - S_0 - V_1) \geq i(G - S_0 - V_1)$$
$$\geq i(G - S_0) + \frac{p_1(G - S_0)}{2}$$
$$> |S_0| \geq 1.$$

According to the choice of V_1 and the definition of $T_{1/2}$, we obtain

$$\omega(G - S_0 \cup (V_1 \setminus x)) = \omega(G - S_0 - (V_1 \setminus x)) = \omega(G - S_0 - V_1)$$
$$\geq |S_0| + 1 \geq 2$$

for $x \in T_{1/2} \subseteq V_1$. From the condition of the theorem and the definition of $\tau(G)$, we obtain

$$2 < \tau(G) \leq \frac{|S_0 \cup (V_1 \setminus x)|}{\omega(G - S_0 \cup (V_1 \setminus x)) - 1} = \frac{|S_0| + |V_1| - 1}{\omega(G - S_0 - V_1) - 1}$$

$$\leq \frac{\omega(G - S_0 - V_1) - 1 + p_1(G - S_0)/2 - 1}{\omega(G - S_0 - V_1) - 1}$$

$$\leq \frac{\omega(G - S_0 - V_1) - 1 + \omega(G - S_0 - V_1) - 1}{\omega(G - S_0 - V_1) - 1}$$

$$= 2,$$

which is a contradiction.

For all cases, we prove that (2.1.37) is not true. Hence, we have

$$2p_0(G - S) + p_1(G - S) \leq 2|S|$$

for any $S \subseteq V(G)$. By Theorem 2.1.2, G contains a fractional 2-factor. □

From Theorem 2.1.16 and Theorem 2.1.17, we pose the following conjecture.

Conjecture 1 [38] Let G be a graph with $\tau(G) > k$, where k is a positive integer. Then G admits a fractional k-factor.

2.2 Fractional k-Factors Including Any Given Edge

In this section, we investigate the existence of fractional k-factors including any given edge in graphs, and obtain two sufficient conditions for graphs to have fractional k-factors including any given edge by using degree and degree sum.

For any $S \subseteq V(G)$ and $T = \{x : x \in V(G) \setminus S, d_{G-S}(x) \leq k\}$, we define $\varepsilon(S, T)$ as follows,

(1) $\varepsilon(S, T) = 2$, if S is not independent.

(2) $\varepsilon(S, T) = 1$, if S is independent and $e_G(S, V(G) \setminus (S \cup T)) \geq 1$, or there exists an edge $e = uv$, such that $u \in S, v \in T$ and $d_{G-S}(v) = k$.

(3) $\varepsilon(S, T) = 0$, if neither (1) nor (2) holds.

Li, Yan and Zhang[10] obtained a necessary and sufficient condition for a graph to have a fractional k-factor including any given edge, which is very

2.2 Fractional k-Factors Including Any Given Edge

useful in the proofs of Theorem 2.2.2 and Theorem 2.2.3.

Theorem 2.2.1 [10] *A graph G has a fractional k-factor including any given edge if and only if for any $S \subseteq V(G)$,*

$$\delta_G(S,T) = k|S| + d_{G-S}(T) - k|T| \geq \varepsilon(S,T),$$

where $T = \{x : x \in V(G) \setminus S, d_{G-S}(x) \leq k\}$ and

$$d_{G-S}(T) = \sum_{x \in T} d_{G-S}(x).$$

Theorem 2.2.2 [41] *Let $k \geq 1$ be an integer and G a graph of order n with $n \geq 4k - 3$, and $\delta(G) \geq k$. If G satisfies*

$$d_G(x) + d_G(y) \geq n + 1 \tag{2.2.1}$$

for each pair of nonadjacent vertices x, y of G, then G has a fractional k-factor including any given edge.

Proof Since $d_G(x) + d_G(y) \geq n + 1$ for each pair of nonadjacent vertices x, y of G, G is Hamiltonian connected. Thus the theorem is true for $k = 1, 2$. In the following we may assume $k \geq 3$. Suppose that there exists an $e \in E(G)$ such that G has no fractional k-factor including e. According to Theorem 2.2.1, there exists a subset S of $V(G)$ such that

$$\delta_G(S,T) = k|S| + d_{G-S}(T) - k|T| \leq \varepsilon(S,T) - 1, \tag{2.2.2}$$

where $T = \{x : x \in V(G) \setminus S, d_{G-S}(x) \leq k\}$.

If $T = \emptyset$, then by (2.2.2),

$$\varepsilon(S,T) - 1 \geq \delta_G(S,T) = k|S| \geq |S| \geq \varepsilon(S,T),$$

a contradiction. Therefore, $T \neq \emptyset$. Define

$$h_1 = \min\{d_{G-S}(x) : x \in T\}.$$

Choose $x_1 \in T$ such that $d_{G-S}(x_1) = h_1$. If $T \setminus N_T[x_1] \neq \emptyset$, let

$$h_2 = \min\{d_{G-S}(x) : x \in T \setminus N_T[x_1]\}.$$

Choose $x_2 \in T \setminus N_T[x_1]$ such that $d_{G-S}(x_2) = h_2$. Thus, we obtain $0 \leq h_1 \leq h_2 \leq k$ by the definition of T and $d_G(x_i) \leq |S| + h_i$ for $i = 1, 2$.

Note that, by (2.2.2),

$$\varepsilon(S,T) - 1 \geq \delta_G(S,T) = k|S| + d_{G-S}(T) - k|T|$$
$$\geq k|S| + h_1|T| - k|T|$$
$$= k|S| - (k - h_1)|T|,$$

that is,
$$\varepsilon(S,T) - 1 \geq k|S| - (k - h_1)|T|. \qquad (2.2.3)$$

Note also that
$$|N_T[x_1]| \leq d_{G-S}(x_1) + 1 = h_1 + 1 \qquad (2.2.4)$$

and $|S| + h_1 \geq d_G(x_1) \geq k$, so that
$$|S| \geq k - h_1. \qquad (2.2.5)$$

We now consider two cases.

Case 1 $T = N_T[x_1]$.

Then $|T| \leq h_1 + 1$ by (2.2.4), and it is easy to see that $\varepsilon(S,T) \leq |S|$ by the definition of $\varepsilon(S,T)$. Thus, using (2.2.3) and (2.2.5) as well,

$$-1 \geq \varepsilon(S,T) - |S| - 1 \geq (k-1)|S| - (k-h_1)|T|$$
$$= (k - h_1)(k - h_1 - 2) + (k - 1)(|S| - k + h_1)$$
$$+ (k - h_1)(h_1 + 1 - |T|)$$
$$\geq -1 + 0 + 0 = -1.$$

Thus equality holds throughout, which implies that $k - h_1 = 1$, $|S| = k - h_1 = 1$ and $|T| = h_1 + 1 = k$. We now consider this case.

Claim 1 $G[T]$ is complete.

Proof If $G[T]$ is not complete, then there exist $u, v \in T$ such that $uv \notin E(G)$. According to the hypothesis of Theorem 2.2.2, we obtain

$$k - 2 \leq n + 1 \leq d_G(u) + d_G(v)$$
$$\leq d_{G-S}(u) + d_{G-S}(v) + 2|S|$$
$$\leq 2k + 2,$$

which implies $k \leq 2$. It contradicts $k \geq 3$. This completes the proof of Claim 1. $\square$

2.2 Fractional k-Factors Including Any Given Edge

According to $|S| = 1$, $|T| = k$, $n \geq 4k - 3$ and $k \geq 3$, we have

$$|V(G) \setminus (S \cup T)| = n - 1 - k \geq 4k - 3 - 1 - k$$
$$= 3k - 4 \geq k + 2.$$

Complying this with Claim 1, $|T| = k$ and $d_{G-S}(x) \leq k$ for any $x \in T$, there exists $t \in V(G) \setminus (S \cup T)$ such that $e_G(t, T) = 0$. Since $x_1 \in T$, we have $x_1 t \notin E(G)$. According to the hypothesis of Theorem 2.2.2, we get

$$n + 1 \leq d_G(x_1) + d_G(t) \leq d_{G-S}(x_1) + |S| + d_G(t)$$
$$\leq k + 1 + (n - k - 1) = n,$$

which is a contradiction.

Case 2 $T \setminus N_T[x_1] \neq \emptyset$.

Obviously, $x_1 x_2 \notin E(G)$. According to the hypothesis of Theorem 2.2.2, we have

$$n + 1 \leq d_G(x_1) + d_G(x_2)$$
$$\leq d_{G-S}(x_1) + d_{G-S}(x_2) + 2|S|$$
$$= h_1 + h_2 + 2|S|,$$

that is,

$$|S| \geq \frac{n + 1 - h_1 - h_2}{2}. \tag{2.2.6}$$

In view of (2.2.4), (2.2.6), $k \geq 3$, $n \geq 4k - 3$, $|S| + |T| \leq n$ and $0 \leq h_1 \leq h_2 \leq k$, we get

$$\delta_G(S, T) = k|S| + d_{G-S}(T) - k|T|$$
$$\geq k|S| + h_1|N_T[x_1]| + h_2(|T| - |N_T[x_1]|) - k|T|$$
$$= k|S| - (h_2 - h_1)|N_T[x_1]| - (k - h_2)|T|$$
$$\geq k|S| - (h_2 - h_1)(h_1 + 1) - (k - h_2)(n - |S|)$$
$$= (2k - h_2)|S| - (h_2 - h_1)(h_1 + 1) - (k - h_2)n$$
$$\geq (2k - h_2) \cdot \frac{n + 1 - h_1 - h_2}{2} - (h_2 - h_1)(h_1 + 1)$$
$$- (k - h_2)n$$

$$= (k-3)(h_2 - h_1) + \frac{h_2(n - 4k + 3 + h_2 - h_1)}{2}$$
$$+ (h_1 - 1)^2 + (k - 1)$$
$$\geq 0 + 0 + 0 + 2 \geq \varepsilon(S, T).$$

This contradicts (2.2.2), and this completes the proof of Theorem 2.2.2. □

Remark 2.2.1 Let us show that the condition (2.2.1) in Theorem 2.2.2 cannot be replaced by
$$d_G(x) + d_G(y) \geq n.$$

Let m and k be two positive integers. We now construct a graph
$$G = (2mk - 1)K_1 \vee K_{2mk-1}.$$
Then $n = |V(G)| = 4mk - 2 > 4k - 3$, $\delta(G) = 2mk - 1 \geq k$ and
$$\min\{d_G(x) + d_G(y) : x, y \in V(G) \text{ and } xy \notin E(G)\}$$
$$= (2mk - 1) + (2mk - 1)$$
$$= 4mk - 2 = n.$$

Clearly, G satisfies all the conditions of Theorem 2.2.2 except the condition (2.2.1). Set $S = V(K_{2mk-1}) \subseteq V(G)$ and
$$T = V((2mk - 1)K_1) \subseteq V(G).$$
Obviously, S is not independent. Therefore, $\varepsilon(S, T) = 2$. Thus, we have
$$\delta_G(S, T) = k|S| + d_{G-S}(T) - k|T|$$
$$= k(2mk - 1) - k(2mk - 1)$$
$$= 0 < 2 = \varepsilon(S, T).$$

By Theorem 2.2.1, G has no fractional k-factor including any given edge. In the above sense, the condition (2.2.1) in Theorem 2.2.2 is best possible.

To replace (2.2.1) with a degree sum condition, Zhou[34] posed a sharp sufficient condition for the existence of fractional k-factors including any given edge in graphs by using the degree.

Theorem 2.2.3 [34] *Let $k \geq 3$ be an integer, and let G be a graph of order n with $n \geq \max\{10, 4k - 3\}$, and $\delta(G) \geq k + 1$. If G satisfies*

2.2 Fractional k-Factors Including Any Given Edge

$$\max\{d_G(x), d_G(y)\} \geq \frac{n+1}{2}$$

for each pair of nonadjacent vertices x, y of G, then G admits a fractional k-factor including any given edge.

Proof Suppose that G satisfies the conditions of Theorem 2.2.3, but it has no fractional k-factor including any given edge. In terms of Theorem 2.2.1, there exists a subset S of $V(G)$ such that

$$\delta_G(S,T) = k|S| + d_{G-S}(T) - k|T| \leq \varepsilon(S,T) - 1, \qquad (2.2.7)$$

where $T = \{x : x \in V(G) \setminus S, d_{G-S}(x) \leq k\}$. In the following we consider three cases.

Case 1 $S = \emptyset$.

In this case, $\varepsilon(S,T) = 0$. In view of (2.2.7), we get

$$-1 \geq \delta_G(S,T) = d_G(T) - k|T| \geq (\delta(G) - k)|T| \geq |T| \geq 0,$$

a contradiction.

Case 2 $|S| = 1$.

In this case, $\varepsilon(S,T) \leq 1$. By (2.2.7) we have

$$0 \geq \delta_G(S,T) = k|S| + d_{G-S}(T) - k|T|$$
$$\geq k|S| + d_G(T) - |T| - k|T|$$
$$= k|S| + d_G(T) - (k+1)|T|$$
$$\geq k|S| + (k+1)|T| - (k+1)|T|$$
$$= k|S| = k \geq 3,$$

this is a contradiction.

Case 3 $|S| \geq 2$.

In this case, $\varepsilon(S,T) \leq 2$. We first prove the following claim.

Claim 1 $|T| \geq k+1$.

Proof If $T = \emptyset$, then by (2.2.7) we have

$$\varepsilon(S,T) - 1 \geq \delta_G(S,T) = k|S| \geq |S| \geq \varepsilon(S,T),$$

which is a contradiction.

If $|T| = 1$, then from (2.2.7) we obtain

$$1 \geq \varepsilon(S,T) - 1 \geq \delta_G(S,T) = k|S| + d_{G-S}(T) - k|T|$$
$$\geq k|S| - k|T| \geq 2k - k = k \geq 3,$$

it is a contradiction.

Hence, $|T| \geq 2$. In the following we assume that $|T| \leq k$. Since $|T| \geq 2$, we have

$$\delta_G(S,T) = k|S| + d_{G-S}(T) - k|T|$$
$$\geq |T||S| + d_{G-S}(T) - k|T|$$
$$= \sum_{x \in T}(|S| + d_{G-S}(x) - k) \geq \sum_{x \in T}(\delta(G) - k)$$
$$= |T| \geq 2 \geq \varepsilon(S,T),$$

which contradicts (2.2.7). This completes the proof of Claim 1. $\square$

According to Claim 1, $T \neq \emptyset$. Define

$$h_1 = \min\{d_{G-S}(x) : x \in T\}.$$

Choose $x_1 \in T$ such that $d_{G-S}(x_1) = h_1$. Furthermore, if $T \setminus N_T[x_1] \neq \emptyset$, we define

$$h_2 = \min\{d_{G-S}(x) : x \in T \setminus N_T[x_1]\}.$$

Choose $x_2 \in T \setminus N_T[x_1]$ such that $d_{G-S}(x_2) = h_2$. Thus, we have

$$0 \leq h_1 \leq h_2 \leq k$$

by the definition of T.

Subcase 3.1 $T = N_T[x_1]$.

From Claim 1 and $T = N_T[x_1]$, we obtain

$$k \geq h_1 = d_{G-S}(x_1) \geq |T| - 1 \geq k.$$

Therefore, $h_1 = k$. According to the definition of h_1, we have

$$\delta_G(S,T) = k|S| + d_{G-S}(T) - k|T| \geq k|S| + h_1|T| - k|T|$$
$$= k|S| + k|T| - k|T| = k|S| \geq |S|$$
$$\geq \varepsilon(S,T).$$

2.2 Fractional k-Factors Including Any Given Edge

That contradicts (2.2.7).

Subcase 3.2 $T \setminus N_T[x_1] \neq \emptyset$.

It is easy to verify that

$$|S| \geq \frac{n+1}{2} - h_2. \qquad (2.2.8)$$

Otherwise, $|S| < \frac{n+1}{2} - h_2$. That is, $|S| + h_2 < \frac{n+1}{2}$, then

$$d_G(x_2) \leq |S| + h_2 < \frac{n+1}{2}$$

and

$$d_G(x_1) \leq |S| + h_1 \leq |S| + h_2 < \frac{n+1}{2}.$$

Since $x_1 x_2 \notin E(G)$, that would contradict the hypothesis of Theorem 2.2.3.

Subcase 3.2.1 $h_2 = 0$.

Clearly, $h_1 = 0$. By (2.2.7), (2.2.8) and $|S| + |T| \leq n$, we obtain

$$\varepsilon(S,T) - 1 \geq \delta_G(S,T) = k|S| + d_{G-S}(T) - k|T|$$
$$\geq k|S| - k|T| \geq k|S| - k(n - |S|)$$
$$= 2k|S| - kn \geq k(n+1) - kn$$
$$= k > 2 \geq \varepsilon(S,T).$$

This is a contradiction.

Subcase 3.2.2 $h_2 \geq 1$.

According to (2.2.8), $|S| + |T| \leq n$, $h_1 \leq h_2 \leq k$ and

$$|N_T[x_1]| \leq d_{G-S}(x_1) + 1 = h_1 + 1,$$

we get

$$\delta_G(S,T) = k|S| + d_{G-S}(T) - k|T|$$
$$\geq k|S| + h_1|N_T[x_1]| + h_2(|T| - |N_T[x_1]|) - k|T|$$
$$= k|S| - (h_2 - h_1)|N_T[x_1]| - (k - h_2)|T|$$
$$\geq k|S| - (h_2 - h_1)(h_1 + 1) - (k - h_2)(n - |S|)$$
$$= (2k - h_2)|S| - (h_2 - h_1)(h_1 + 1) - (k - h_2)n$$
$$\geq (2k - h_2)\left(\frac{n+1}{2} - h_2\right) - (h_2 - h_1)(h_1 + 1) - (k - h_2)n$$

55

$$= h_2^2 + \left(\frac{n}{2} - 2k - \frac{3}{2}\right)h_2 - h_1h_2 + h_1^2 + h_1 + k$$

$$= \left(\frac{h_2-1}{2} - h_1\right)^2 + \frac{3}{4}h_2^2 + \left(\frac{n}{2} - 2k - 1\right)h_2 + k - \frac{1}{4}$$

$$\geq \frac{3}{4}h_2^2 + \left(\frac{n}{2} - 2k - 1\right)h_2 + k - \frac{1}{4},$$

that is,

$$\delta_G(S,T) \geq \frac{3}{4}h_2^2 + \left(\frac{n}{2} - 2k - 1\right)h_2 + k - \frac{1}{4}. \qquad (2.2.9)$$

If $k = 3$, then $n \geq 10$. Hence, we have by (2.2.9)

$$\delta_G(S,T) \geq \frac{3}{4}h_2^2 - 2h_2 + 3 - \frac{1}{4} > 1.$$

In view of the integrity of $\delta_G(S,T)$, we obtain

$$\delta_G(S,T) \geq 2 \geq \varepsilon(S,T).$$

This contradicts (2.2.7).

If $k \geq 4$, then $n \geq 4k - 3$. Therefore, from (2.2.9) we get

$$\delta_G(S,T) \geq \frac{3}{4}h_2^2 + \left(\frac{4k-3}{2} - 2k - 1\right)h_2 + k - \frac{1}{4}$$

$$\geq \frac{3}{4}h_2^2 - \frac{5}{2}h_2 + 4 - \frac{1}{4} > 1.$$

According to the integrity of $\delta_G(S,T)$, we have

$$\delta_G(S,T) \geq 2 \geq \varepsilon(S,T).$$

Which contradicts (2.2.7). This completes the proof of Theorem 2.2.3. $\square$

Remark 2.2.2 Let us show that the condition

$$\max\{d_G(x), d_G(y)\} \geq \frac{n+1}{2}$$

in Theorem 2.2.3 cannot be replaced by

$$\max\{d_G(x), d_G(y)\} \geq \frac{n}{2}.$$

Let $t \geq 2$ and $k \geq 3$ be two integers. We construct a graph

$$G = ((kt-2)K_1 \cup K_2) \vee ktK_1.$$

Clearly, $\delta(G) = kt \geq 2k > k+1$, $n = |V(G)| = 2kt \geq 4k > 4k - 3$ and

$$\max\{d_G(x), d_G(y)\} = \frac{n}{2}$$

for each pair of nonadjacent vertices x, y of $((kt-2)K_1 \cup ktK_1) \subset G$. Let

$$S = V((kt-2)K_1 \cup K_2)) \subseteq V(G)$$

and $T = V(ktK_1) \subseteq V(G)$. Then $|S| = kt$, $|T| = kt$ and S is not independent. Thus, we get $\varepsilon(S, T) = 2$ and

$$\delta_G(S, T) = k|S| + d_{G-S}(T) - k|T| = k^2 t - k^2 t$$
$$= 0 < 2 = \varepsilon(S, T).$$

According to Theorem 2.2.1, G has no fractional k-factor including any given edge. In the above sense, the result in Theorem 2.2.3 is best possible.

2.3 Fractional (g, f)-Factors with Prescribed Properties

In the previous sections, we discussed the existence of fractional k-factors in graphs. In this section, we extend the discussion to more general fractional factors, fractional (g, f)-factors. Furthermore, we present two sufficient conditions for the existence of fractional (g, f)-factors with prescribed properties in graphs.

We shall show the fractional (g, f)-factor theorem presented by Anstee[2]. Liu and Zhang[12] gave a simple proof of the theorem.

Theorem 2.3.1 [2, 12] *Let G be a graph. Then G has a fractional (g, f)-factor if and only if for every subset S of $V(G)$,*

$$\delta_G(S, T) = f(S) + d_{G-S}(T) - g(T) \geq 0,$$

where $T = \{x : x \in V(G) \setminus S, \, d_{G-S}(x) \leq g(x)\}$.

Theorem 2.3.2 [4] *Let G be a graph, and let a, b and r be three nonnegative integers satisfying $1 \leq a \leq b - r$, and let g, f be two integer-valued functions defined on $V(G)$ with $a \leq g(x) \leq f(x) - r \leq b - r$ for every*

$x \in V(G)$. If
$$\kappa(G) \geq \max\left\{\frac{(b+1)(b-r+1)}{2}, \frac{(b-r+1)^2 \alpha(G)}{4(a+r)}\right\},$$
then G contains a fractional (g, f)-factor.

Proof We prove Theorem 2.3.2 by contradiction. Suppose that G satisfies the assumption of Theorem 2.3.2, but it has no fractional (g, f)-factor. Then using Theorem 2.3.1, there exists some subset S of $V(G)$ satisfying
$$\delta_G(S, T) = f(S) + d_{G-S}(T) - g(T) \leq -1, \qquad (2.3.1)$$
where $T = \{x : x \in V(G) \setminus S, d_{G-S}(x) \leq g(x)\}$. Obviously, $T \neq \emptyset$ by (2.3.1).

We present the following partition of T: we choose $x_1 \in T$ with
$$d_{G[T]}(x_1) = \delta(G[T]).$$
Let $D_1 = N_G[x_1] \cap T$ and $T_1 = T$. If $T - \bigcup_{1 \leq j < i} D_j \neq \emptyset$ for $i \geq 2$, then we write
$$T_i = T - \bigcup_{1 \leq j < i} D_j.$$
In the following, we take $x_i \in T_i$ with $d_{G[T_i]}(x_i) = \delta(G[T_i])$ and $D_i = N_G[x_i] \cap T_i$. We continue these procedures until we reach the situation in which $T_i = \emptyset$ for some i, say for $i = s + 1$. It follows from the above definition that $\{x_1, x_2, \cdots, x_s\}$ is an independent set of G.

Note that $T \neq \emptyset$. Hence, we have $s \geq 1$. Set $|D_i| = d_i$, we obtain
$$|T| = \sum_{1 \leq i \leq s} d_i.$$
We write $U = V(G) \setminus (S \cup T)$ and $\kappa(G - S) = t$. Now, we verify the following claims.

Claim 1 $s \neq 1$ or $U \neq \emptyset$.

Proof Assume that $s = 1$ and $U = \emptyset$. Then by (2.3.1) and our choice of x_1, we obtain
$$-1 \geq \delta_G(S, T) = f(S) + d_{G-S}(T) - g(T)$$

2.3 Fractional (g, f)-Factors with Prescribed Properties

$$\geq (a+r)|S| + d_{G-S}(T) - (b-r)|T|$$
$$= (a+r)|S| + d_1(d_1 - 1) - (b-r)d_1,$$

that is,

$$|S| \leq \frac{-d_1^2 + (b-r+1)d_1 - 1}{a+r}. \qquad (2.3.2)$$

In terms of (2.3.2), we have

$$|V(G)| = |S| + d_1 \leq \frac{-d_1^2 + (b-r+1)d_1 - 1}{a+r} + d_1$$
$$= \frac{-d_1^2 + (a+b+1)d_1 - 1}{a+r} \leq \frac{(a+b+1)^2 - 4}{4(a+r)}$$
$$\leq \frac{(b+1)(b-r+1)}{2}.$$

Combining this with the hypothesis of Theorem 2.3.2, we obtain

$$\frac{(b+1)(b-r+1)}{2} \geq |V(G)| > \kappa(G) \geq \frac{(b+1)(b-r+1)}{2},$$

which is a contradiction. The proof of Claim 1 is complete. □

Claim 2 $d_{G-S}(T) \geq \sum_{1 \leq i \leq s} d_i(d_i - 1) + \dfrac{st}{2}.$

Proof According to the choice of x_i, we have

$$\sum_{1 \leq i \leq s} \Big(\sum_{x \in D_i} d_{G[T_i]}(x)\Big) \geq \sum_{1 \leq i \leq s} d_i(d_i - 1). \qquad (2.3.3)$$

For the left-hand side of (2.3.3), an edge joining $x \in D_i$ and $y \in D_j$ ($i < j$) is counted only once, that is to say, it is counted in $d_{G[T_i]}(x)$ but not in $d_{G[T_j]}(y)$. Thus, we obtain

$$d_{G-S}(T) \geq \sum_{1 \leq i \leq s} d_i(d_i - 1) + \sum_{1 \leq i < j \leq s} e_G(D_i, D_j) + e_G(T, U).$$

$$(2.3.4)$$

According to $\kappa(G - S) = t$ and Claim 1, we have

$$e_G\Big(D_i, \bigcup_{j \neq i} D_j\Big) + e_G(D_i, U) \geq t \qquad (2.3.5)$$

for all D_i ($1 \leq i \leq s$). It follows from (2.3.5) that

$$\sum_{1 \leq i \leq s} \left(e_G\left(D_i, \bigcup_{j \neq i} D_j\right) + e_G(D_i, U) \right)$$

$$= 2 \sum_{1 \leq i < j \leq s} e_G(D_i, D_j) + e_G(T, U) \geq st,$$

which implies

$$\sum_{1 \leq i < j \leq s} e_G(D_i, D_j) + e_G(T, U) \geq \frac{st}{2}.$$

Combining this with (2.3.4), we obtain

$$d_{G-S}(T) \geq \sum_{1 \leq i \leq s} d_i(d_i - 1) + \frac{st}{2}.$$

This completes the proof of Claim 2. $\square$

It is easy to see that

$$d_i^2 - (b - r + 1)d_i \geq -\frac{(b - r + 1)^2}{4}.$$

Combining this with $|T| = \sum_{1 \leq i \leq s} d_i$ and Claim 2, we have

$$\delta_G(S, T) = f(S) + d_{G-S}(T) - g(T)$$

$$\geq (a + r)|S| + \sum_{1 \leq i \leq s} d_i(d_i - 1) + \frac{st}{2} - (b - r)|T|$$

$$= (a + r)|S| + \sum_{1 \leq i \leq s} d_i(d_i - 1) + \frac{st}{2} - (b - r) \sum_{1 \leq i \leq s} d_i$$

$$= (a + r)|S| + \sum_{1 \leq i \leq s} (d_i^2 - (b - r + 1)d_i) + \frac{st}{2}$$

$$\geq (a + r)|S| - \frac{(b - r + 1)^2 s}{4} + \frac{st}{2},$$

that is,

$$\delta_G(S, T) \geq (a + r)|S| - \frac{(b - r + 1)^2 s}{4} + \frac{st}{2}. \quad (2.3.6)$$

Claim 3 $-\frac{(b - r + 1)^2}{4} + \frac{t}{2} < 0.$

Proof If $-\frac{(b-r+1)^2}{4} + \frac{t}{2} \geq 0$, then by (2.3.6), $s \geq 1$ and $|S| \geq 0$, we

2.3 Fractional (g, f)-Factors with Prescribed Properties

obtain
$$\delta_G(S, T) \geq (a + r)|S| - \frac{(b - r + 1)^2 s}{4} + \frac{st}{2} \geq 0,$$

which contradicts (2.3.1). The proof of Claim 3 is complete. □

Note that $\alpha(G) \geq \alpha(G[T]) \geq s$ and
$$\kappa(G) \leq |S| + \kappa(G - S) = |S| + t.$$

Combining these with (2.3.1), (2.3.6), Claim 3 and the condition
$$\kappa(G) \geq \max \left\{ \frac{(b+1)(b-r+1)}{2}, \frac{(b-r+1)^2 \alpha(G)}{4(a+r)} \right\}$$

of Theorem 2.3.2, we have
$$-1 \geq \delta_G(S, T) \geq (a + r)|S| - \frac{(b - r + 1)^2 s}{4} + \frac{st}{2}$$
$$= (a + r)|S| + \left(-\frac{(b - r + 1)^2}{4} + \frac{t}{2} \right) s$$
$$\geq (a + r)(\kappa(G) - t) + \left(-\frac{(b - r + 1)^2}{4} + \frac{t}{2} \right) \alpha(G)$$
$$\geq (a + r)(\kappa(G) - t) + \left(-\frac{(b - r + 1)^2}{4} + \frac{t}{2} \right) \cdot \frac{4(a+r)\kappa(G)}{(b-r+1)^2}$$
$$= (a + r)t \left(\frac{2\kappa(G)}{(b - r + 1)^2} - 1 \right) \geq 0,$$

which is a contradiction. This completes the proof of Theorem 2.3.2. □

Remark 2.3.1 We show that the condition
$$\kappa(G) \geq \frac{(b - r + 1)^2 \alpha(G)}{4(a + r)} = \frac{\left(\frac{b-r+1}{2}\right)^2 \alpha(G)}{a + r}$$

in Theorem 2.3.2 is sharp by constructing a graph
$$G = K_{\left(\frac{b-r+1}{2}\right)^2 s - 1 \atop a + r} \vee \left(s K_{\frac{b-r+1}{2}} \right),$$

where a, b, r are three nonnegative integers with $1 \leq a = b - r$, s is a sufficiently large integer, $\frac{b-r+1}{2}$ and $\frac{\left(\frac{b-r+1}{2}\right)^2 s - 1}{a+r}$ are two integers. It is easy to see that $\alpha(G) = s$ and

$$\kappa(G) = \frac{\left(\frac{b-r+1}{2}\right)^2 s - 1}{a+r} = \frac{\left(\frac{b-r+1}{2}\right)^2 \alpha(G) - 1}{a+r}.$$

Let g and f be two functions defined on $V(G)$ with $g(x) \equiv a$ and $f(x) \equiv b$. In the following, we prove that G has no fractional (g, f)-factor.

We take $S = V\left(K_{\left(\frac{b-r+1}{2}\right)^2 s - 1 \atop a+r}\right)$ and $T = V\left(sK_{\frac{b-r+1}{2}}\right)$. Note that $a = b - r$. Thus, we have

$$\delta_G(S, T) = f(S) + d_{G-S}(T) - g(T) = b|S| + d_{G-S}(T) - a|T|$$

$$= b \cdot \frac{\left(\frac{b-r+1}{2}\right)^2 s - 1}{a+r} + \frac{(b-r+1)s}{2} \cdot \left(\frac{b-r+1}{2} - 1\right)$$

$$- a \cdot \frac{(b-r+1)s}{2}$$

$$= \left(\frac{b-r+1}{2}\right)^2 s - 1 - \frac{(b-r+1)s}{2} \cdot \left(b - r - \frac{b-r+1}{2} + 1\right)$$

$$= -1 < 0.$$

Then using Theorem 2.3.1, G has no fractional (g, f)-factor.

If $r = 0$ in Theorem 2.3.2, then we obtain the following corollary.

Corollary 1 *Let G be a graph, and let a, b be two integers satisfying $1 \leq a \leq b$, and let g, f be two integer-valued functions defined on $V(G)$ with $a \leq g(x) \leq f(x) \leq b$ for every $x \in V(G)$. If*

$$\kappa(G) \geq \max\left\{\frac{(b+1)^2}{2}, \frac{(b+1)^2 \alpha(G)}{4a}\right\},$$

then G admits a fractional (g, f)-factor.

If $g(x) \equiv f(x)$ in Corollary 1, then we get the following corollary.

Corollary 2 *Let G be a graph, and let a, b be two integers satisfying $1 \leq a \leq b$, and let f be an integer-valued function defined on $V(G)$ with $a \leq f(x) \leq b$ for every $x \in V(G)$. If*

$$\kappa(G) \geq \max\left\{\frac{(b+1)^2}{2}, \frac{(b+1)^2 \alpha(G)}{4a}\right\},$$

2.3 Fractional (g, f)-Factors with Prescribed Properties

then G have a fractional f-factor.

If $a = b = k$ in Corollary 2, then we have the following corollary.

Corollary 3 Let G be a graph, and let k be an integer with $k \geq 1$. If
$$\kappa(G) \geq \max\left\{\frac{(k+1)^2}{2}, \frac{(k+1)^2\alpha(G)}{4k}\right\},$$
then G have a fractional k-factor.

Liu and Zhang[13] obtained a sufficient condition for a graph to have a fractional (g, f)-factor.

Theorem 2.3.3 [13] Let G be a graph, and let g, f be two nonnegative integer-valued functions defined on $V(G)$ with $g(x) \leq f(x)$ for every $x \in V(G)$. If
$$f(x)d_G(y) \geq d_G(x)g(y)$$
for any $x, y \in V(G)$ with $x \neq y$, then G admits a fractional (g, f)-factor.

Yang, Ma and Liu[22] proved the following theorem, which is an extension of Theorem 2.3.3.

Theorem 2.3.4 [22] Let G be a graph, and let g, f be two nonnegative integer-valued functions defined on $V(G)$ with $g(x) \leq f(x) \leq d_G(x)$ for every $x \in V(G)$. If
$$(f(x) - 1)d_G(y) \geq (d_G(x) - 1)g(y)$$
for any $x, y \in V(G)$ with $x \neq y$, then G has a fractional (g, f)-factor including any given edge.

Zhou and Shang[43] showed the existence of fractional (g, f)-factors with prescribed properties in graphs, which is an extension of Theorem 2.3.3 and Theorem 2.3.4.

Theorem 2.3.5 [43] Let G be a graph, and let g, f be two nonnegative integer-valued functions defined on $V(G)$ with $g(x) \leq f(x) \leq d_G(x)$ for

every $x \in V(G)$. If
$$(f(x) - k)d_G(y) \geq (d_G(x) - k)g(y)$$
for any $x, y \in V(G)$ with $x \neq y$, then G has a fractional (g, f)-factor including any given k edges.

Proof We apply induction on k. The theorem is true for $k = 1$ by Theorem 2.3.4. Suppose the theorem holds for $k = n$. In the following, we prove that the theorem holds for $k = n + 1$.

According to the condition of the theorem, we have
$$(f(x) - (n+1))d_G(y) \geq (d_G(x) - (n+1))g(y).$$

Let $e = uv \in E(G)$, $G' = G - e$. Two integer-valued functions $g'(x)$ and $f'(x)$ defined on $V(G)$ are defined as follows:

$$g'(x) = \begin{cases} g(x) - 1, & x = u, v, \\ g(x), & x \neq u, v, \end{cases}$$

$$f'(x) = \begin{cases} f(x) - 1, & x = u, v, \\ f(x), & x \neq u, v. \end{cases}$$

Next we prove that G' has a fractional (g', f')-factor including any given n edges. By the hypothesis, we only need prove
$$(f'(x) - n)d_{G'}(y) \geq (d_{G'}(x) - n)g'(y) \qquad (2.3.7)$$
for any $x, y \in V(G')$ with $x \neq y$. We consider four cases.

Case 1 $x \notin \{u, v\}$, $y \notin \{u, v\}$.

In this case, we have $d_G(x) = d_{G'}(x)$, $f(x) = f'(x)$, $d_G(y) = d_{G'}(y)$ and $g(y) = g'(y)$. Thus,

$$\begin{aligned}(f'(x) - n)d_{G'}(y) &= (f(x) - (n+1))d_G(y) + d_G(y) \\ &\geq (d_G(x) - (n+1))g(y) + d_G(y) \\ &= (d_G(x) - n)g(y) + d_G(y) - g(y) \\ &\geq (d_G(x) - n)g(y) \\ &= (d_{G'}(x) - n)g'(y).\end{aligned}$$

2.3 Fractional (g,f)-Factors with Prescribed Properties

Case 2 $x \in \{u,v\}$, $y \notin \{u,v\}$.

Obviously, $d_{G'}(x) = d_G(x) - 1$, $f'(x) = f(x) - 1$, $d_{G'}(y) = d_G(y)$ and $g'(y) = g(y)$. Thus, we obtain

$$(f'(x) - n)d_{G'}(y) = (f(x) - (n+1))d_G(y) \geq (d_G(x) - (n+1))g(y)$$
$$= (d_{G'}(x) - n)g'(y).$$

Case 3 $x \notin \{u,v\}$, $y \in \{u,v\}$.

Clearly, $d_{G'}(x) = d_G(x)$, $f'(x) = f(x)$, $d_{G'}(y) = d_G(y) - 1$ and $g'(y) = g(y) - 1$. Hence, we have

$$(f'(x) - n)d_{G'}(y) = (f(x) - (n+1))d_G(y) + d_G(y) - 1 - f(x) + (n+1)$$
$$\geq (d_G(x) - (n+1))g(y) + d_G(y) - f(x) + n$$
$$= (d_{G'}(x) - n)g'(y) + d_G(x) - n - f(x) + n$$
$$\quad + d_G(y) - g(y)$$
$$\geq (d_{G'}(x) - n)g'(y).$$

Case 4 $x \in \{u,v\}$, $y \in \{u,v\}$.

It is obvious that $d_{G'}(x) = d_G(x) - 1$, $f'(x) = f(x) - 1$, $d_{G'}(y) = d_G(y) - 1$ and $g'(y) = g(y) - 1$. Therefore, we obtain

$$(f'(x) - n)d_{G'}(y) = (f(x) - (n+1))(d_G(y) - 1)$$
$$= (f(x) - (n+1))d_G(y) - f(x) + (n+1)$$
$$\geq (d_G(x) - (n+1))g(y) - f(x) + (n+1)$$
$$= (d_{G'}(x) - n)g'(y) + d_G(x) - (n+1)$$
$$\quad - f(x) + (n+1)$$
$$\geq (d_{G'}(x) - n)g'(y).$$

We prove that (2.3.7) holds for any $x, y \in V(G')$ with $x \neq y$. By hypothesis, G' has a fractional (g', f')-factor including any given n edges, that is, G has a fractional (g, f)-factor including any given $n+1$ edges.

By the induction hypothesis, G has a fractional (g, f)-factor including any given k edges. This completes the proof of Theorem 2.3.5. □

Chapter 3
Fractional Deleted Graphs

In this chapter, we discuss a generalization of fractional factors, i.e., fractional deleted graphs from different perspectives, such as degree condition, neighborhood condition and binding number. We present some sufficient conditions related to these parameters for the existence of fractional k-deleted graphs and fractional (g, f)-deleted graphs. Furthermore, it is shown that these results are sharp.

3.1 Fractional k-Deleted Graphs

Let G be a graph and k a positive integer. A graph G is a k-deleted graph if there exists a k-factor excluding any given edge in G. A graph G is a fractional k-deleted graph if there exists a fractional k-factor after deleting any edge of G. A necessary and sufficient condition for a graph to be a fractional k-deleted graph was obtained by Li, Yan and Zhang[10] in 2003. In this section, we use the result to discuss the relationship between graphic parameters and fractional k-deleted graphs, and pose some sufficient conditions for graphs to be fractional k-deleted graphs. Furthermore, we show the results are best possible in some sense.

Theorem 3.1.1 [10] *A graph G is a fractional k-deleted graph if and only if for any $S \subseteq V(G)$ and $T = \{x : x \in V(G) \setminus S, d_{G-S}(x) \leq k\}$,*

$$\delta_G(S,T) = k|S| + d_{G-S}(T) - k|T| \geq \varepsilon(S,T),$$

or

3.1 Fractional k-Deleted Graphs

$$k|S| - \sum_{j=0}^{k-1}(k-j)p_j(G-S) \geq \varepsilon(S,T),$$

where $p_j(G-S) = |\{x : d_{G-S}(x) = j\}|$, and $\varepsilon(S,T)$ is defined as follows,

$$\varepsilon(S,T) = \begin{cases} 2, & \text{if } T \text{ is not independent,} \\ 1, & \text{if } T \text{ is independent, and } e_G(T, V(G) \setminus (S \cup T)) \geq 1, \\ 0, & \text{otherwise.} \end{cases}$$

We now show some degree conditions for graphs to be fractional k-deleted graphs.[33]

Theorem 3.1.2 [33] *Let G be a 2-connected graph of order n with $n \geq 7$ and $\delta(G) \geq 2$. If*

$$\max\{d_G(x), d_G(y)\} \geq \frac{n}{2}$$

for each pair of nonadjacent vertices x, y of G, then G is a fractional 1-deleted graph.

Proof Suppose that G satisfies the assumption of the theorem, but it is not a fractional 1-deleted graph. Then by Theorem 3.1.1, there exists some $S \subseteq V(G)$ such that

$$i(G-S) > |S| - \varepsilon(S,T), \tag{3.1.1}$$

where $T = \{x : x \in V(G) \setminus S, d_{G-S}(x) \leq 1\}$. Clearly, $S \neq \emptyset$ since $\delta(G) \geq 2$ and G is 2-connected.

Claim 1 $|T| > |S|$.

Proof If T is not independent, then $\varepsilon(S,T) = 2$ and $|T| \geq i(G-S) + 2$. According to (5.2.1), we obtain

$$|T| \geq i(G-S) + 2 > |S| - \varepsilon(S,T) + 2 = |S|.$$

If T is independent and $e_G(T, V(G) \setminus (S \cup T)) \geq 1$, then $\varepsilon(S,T) = 1$ and $|T| \geq i(G-S) + 1$. In view of (5.2.1), we get

$$|T| \geq i(G-S) + 1 > |S| - \varepsilon(S,T) + 1 = |S|.$$

Otherwise, $\varepsilon(S,T) = 0$ and $|T| \geq i(G-S)$. From (5.2.1), we have

$$|T| \geq i(G-S) > |S| - \varepsilon(S,T) = |S|.$$

This completes the proof of Claim 1. □

According to Claim 1 and $|S| + |T| \leq n$, we have

$$|S| < \frac{n}{2}. \tag{3.1.2}$$

Case 1 $|S| = 1$.

In view of Claim 1, we have $|T| \geq 2$.

Subcase 1.1 T is not independent.

In this case, $\varepsilon(S,T) = 2$. We prove firstly the following claim.

Claim 2 $|T| \geq 3$.

Proof If $T = 2$, then $e = uv \in E(G[T])$ for $u, v \in T$. Since G is 2-connected, $G - S$ is connected. From $n \geq 7$, then there exists $x \in V(G) \setminus (S \cup T)$ such that x is adjacent to u or v. Hence, $d_{G-S}(u) \geq 2$ or $d_{G-S}(v) \geq 2$, which contradicts $d_{G-S}(y) \leq 1$ for any $y \in T$. This completes the proof of Claim 2. □

According to $|T| \geq 3$ and the definition of T, then there exist $u, v \in T$ such that $uv \notin E(G)$. By the hypothesis of Theorem 3.1.2, we obtain

$$\frac{n}{2} \leq \max\{d_G(u), d_G(v)\} \leq \max\{d_{G-S}(u), d_{G-S}(v)\} + |S|$$

$$\leq |S| + 1 = 2,$$

which implies $n \leq 4$. This contradicts $n \geq 7$.

Subcase 1.2 T is independent.

In this case, $\varepsilon(S,T) \leq 1$. Since G is 2-connected, then $G - S$ is connected. Thus, we have

$$i(G-S) = 0. \tag{3.1.3}$$

On the other hand, from (5.2.1) we obtain

$$i(G-S) > |S| - \varepsilon(S,T) \geq |S| - 1 = 0,$$

which contradicts (5.2.3).

Case 2 $|S| = 2$.

By Claim 1, we have $|T| \geq 3$. According to (5.2.1) and $\varepsilon(S,T) \leq 2$, we

obtain
$$i(G-S) > |S| - \varepsilon(S,T) \geq |S| - 2 = 0.$$

Therefore, there exists $x \in T$ with $d_{G-S}(x) = 0$. Since $|T| \geq 3$, we have $y \in T \setminus \{x\}$ such that $d_{G-S}(y) \leq 1$ and $xy \notin E(G)$. In view of $n \geq 7$ and the hypothesis of Theorem 3.1.2, we have
$$\frac{7}{2} \leq \frac{n}{2} \leq \max\{d_G(x), d_G(y)\} \leq d_{G-S}(y) + |S| \leq 3,$$

which is a contradiction.

Case 3 $|S| \geq 3$.

According to $\varepsilon(S,T) \leq 2$ and (5.2.1), we get that
$$i(G-S) \geq |S| - \varepsilon(S,T) + 1 \geq |S| - 1 \geq 2.$$

Let $I(G-S) = \{x_1, x_2, \cdots, x_{i(G-S)}\}$. Clearly, $x_i x_j \notin E(G)$ ($i \neq j$ and $i, j \in \{1, 2, \cdots, i(G-S)\}$). Hence, for any two vertices x_i and x_j, we obtain
$$\max\{d_G(x_i), d_G(x_j)\} \geq \frac{n}{2}$$

by the hypothesis of Theorem 3.1.2. But by (5.2.2),
$$d_G(x_i) \leq |S| < \frac{n}{2}$$

for all $x_i \in I(G-S)$. This is a contradiction. This completes the proof of Theorem 3.1.2. □

Theorem 3.1.3 [33] *Let G be a 2-connected graph of order n with $n \geq 9$ and $\delta(G) \geq 3$. If*
$$\max\{d_G(x), d_G(y)\} \geq \frac{n}{2}$$

for each pair of nonadjacent vertices x, y of G, then G is a fractional 2-deleted graph.

Proof Suppose that G satisfies the assumption of Theorem 3.1.3, but it is not a fractional 2-deleted graph. Then by Theorem 3.1.1, there exists some $S \subseteq V(G)$ such that
$$2|S| - 2p_0(G-S) - p_1(G-S) \leq \varepsilon(S,T) - 1, \qquad (3.1.4)$$

where $T = \{x : x \in V(G) \setminus S, d_{G-S}(x) \leq 2\}$. If $S = \emptyset$, then we have $p_0(G-S) = p_1(G-S) = 0$ by $\delta(G) \geq 3$. In this case, it is easy to see that $\varepsilon(S,T) = 0$. From (3.1.4), we obtain

$$0 = 2|S| - 2p_0(G-S) - p_1(G-S) \leq \varepsilon(S,T) - 1 = -1,$$

this is a contradiction. If $|S| = 1$, then we have $p_0(G-S) = p_1(G-S) = 0$ by $\delta(G) \geq 3$. According to (3.1.4) and $\varepsilon(S,T) \leq 2$, we get

$$2 = 2|S| = 2|S| - 2p_0(G-S) - p_1(G-S)$$
$$\leq \varepsilon(S,T) - 1 \leq 1,$$

a contradiction. In the following we may assume that $|S| \geq 2$. We prove firstly the following claim.

Claim 3 $|T| > |S|$.

Proof If T is not independent, then $\varepsilon(S,T) = 2$. Let $p_1(G-S) \geq 2$. Then

$$|T| \geq p_0(G-S) + p_1(G-S).$$

Complying this with (3.1.4), we have

$$|T| \geq p_0(G-S) + p_1(G-S)$$
$$\geq |S| + \frac{p_1(G-S) - \varepsilon(S,T) + 1}{2}$$
$$\geq |S| + \frac{1}{2}.$$

According to the integrity of $|T|$ and $|S|$, we have $|T| > |S|$. Let $p_1(G-S) \leq 1$. Then $|T| \geq p_0(G-S) + 2$. In view of (3.1.4), we obtain

$$|T| \geq p_0(G-S) + 2 \geq p_0(G-S) + p_1(G-S) + 1$$
$$\geq |S| + \frac{p_1(G-S) - \varepsilon(S,T) + 1}{2} + 1$$
$$\geq |S| + \frac{1}{2}.$$

By the integrity of $|T|$ and $|S|$, we have $|T| > |S|$.

If T is independent and $e_G(T, V(G) \setminus (S \cup T)) \geq 1$, then $\varepsilon(S,T) = 1$. Let $p_1(G-S) \geq 1$. Then we have $|T| \geq p_0(G-S) + p_1(G-S)$. By (3.1.4),

3.1 Fractional k-Deleted Graphs

$$|T| \geq p_0(G-S) + p_1(G-S)$$
$$\geq |S| + \frac{p_1(G-S) - \varepsilon(S,T) + 1}{2}$$
$$\geq |S| + \frac{1}{2}.$$

In view of the integrity of $|T|$ and $|S|$, we obtain $|T| > |S|$. Let $p_1(G-S) = 0$. From T is independent and $e_G(T, V(G) \setminus (S \cup T)) \geq 1$, we get $|T| > p_0(G-S)$. By $p_1(G-S) = 0$, $\varepsilon(S,T) = 1$ and (3.1.4),

$$p_0(G-S) \geq |S|. \qquad (3.1.5)$$

From $|T| > p_0(G-S)$ and (3.1.5), we have

$$|T| > p_0(G-S) \geq |S|.$$

Otherwise, $\varepsilon(S,T) = 0$ and $|T| \geq p_0(G-S) + p_1(G-S)$. Complying this with (3.1.4), we obtain

$$|T| \geq p_0(G-S) + p_1(G-S) \geq |S| + \frac{1}{2}.$$

According to the integrity of $|T|$ and $|S|$, we obtain $|T| > |S|$. This completes the proof of Claim 3. □

In view of Claim 3 and $|S| + |T| \leq n$, we have

$$|S| < \frac{n}{2}. \qquad (3.1.6)$$

Case 1 $p_0(G-S) \geq 2$.

Obviously, $|I(G-S)| = p_0(G-S) \geq 2$. For any $x, y \in I(G-S)$, we have $xy \notin E(G)$. According to the hypothesis of Theorem 3.1.3, there must exist at least one vertex, say y, such that

$$d_G(y) \geq \frac{n}{2}.$$

Complying this with $d_{G-S}(y) = 0$ and (3.1.6), we obtain

$$\frac{n}{2} \leq d_G(y) \leq d_{G-S}(y) + |S| = |S| < \frac{n}{2}.$$

This is a contradiction.

Case 2 $p_0(G-S) = 1$.

Subcase 2.1 $p_1(G - S) = 0$.

According to (3.1.4) and $\varepsilon(S, T) \leq 2$, we have

$$2|S| \leq 2p_0(G - S) + p_1(G - S) + 1 = 3,$$

which contradicts $|S| \geq 2$.

Subcase 2.2 $p_1(G - S) \geq 1$.

Since $p_0(G - S) = 1$ and $p_1(G - S) \geq 1$, there exist $x, y \in T$ such that $d_{G-S}(x) = 0$ and $d_{G-S}(y) = 1$. Obviously, $xy \notin E(G)$. According to the hypothesis of Theorem 3.1.3, we obtain

$$\frac{n}{2} \leq \max\{d_G(x), d_G(y)\} \leq d_{G-S}(y) + |S| = |S| + 1,$$

that is,

$$|S| \geq \frac{n}{2} - 1. \tag{3.1.7}$$

If n is odd and since $|S|$ is an integer, we have $|S| \geq \frac{n-1}{2}$. Moreover,

$$p_1(G - S) \leq n - p_0(G - S) - |S| \leq n - 1 - \frac{n-1}{2} = \frac{n-1}{2}.$$

Hence, by (3.1.4) we obtain

$$1 \geq \varepsilon(S, T) - 1 \geq 2|S| - 2p_0(G - S) - p_1(G - S)$$

$$\geq 2 \cdot \frac{n-1}{2} - 2 - \frac{n-1}{2} = \frac{n-5}{2},$$

which implies $n \leq 7$. This contradicts $n \geq 9$.

If n is even, then by (3.1.6) and (3.1.7) we have $|S| = \frac{n}{2} - 1$. Furthermore,

$$p_1(G - S) \leq n - p_0(G - S) - |S| = n - 1 - \left(\frac{n}{2} - 1\right) = \frac{n}{2}.$$

Since $n \geq 9$ and n is even, we have $n \geq 10$.

For $n = 10$, we have $p_1(G - S) \neq \frac{n}{2} = 5$. Otherwise,

$$n = |S| + p_0(G - S) + p_1(G - S).$$

We write $P_1(G - S) = \{x : x \in V(G) \setminus S, d_{G-S}(x) = 1\}$ and

$$p_1(G - S) = |P_1(G - S)|.$$

3.1 Fractional k-Deleted Graphs

Then for each $x \in P_1(G-S)$, x is only adjacent to the vertices of $P_1(G-S)$ in $G-S$. It is impossible by $p_1(G-S) = 5$ and the definition of $P_1(G-S)$. Thus, we obtain $p_1(G-S) \leq 4$. Complying this with (3.1.4) and $\varepsilon(S,T) \leq 2$,

$$1 \geq \varepsilon(S,T) - 1 \geq 2|S| - 2p_0(G-S) - p_1(G-S)$$
$$\geq 8 - 2 - 4 = 2,$$

it is a contradiction.

For $n \geq 12$, we obtain by $|S| = \frac{n}{2} - 1$ and $p_1(G-S) \leq \frac{n}{2}$,

$$2|S| - 2p_0(G-S) - p_1(G-S)$$
$$\geq 2\left(\frac{n}{2} - 1\right) - 2 - \frac{n}{2} = \frac{n}{2} - 4$$
$$\geq 2 \geq \varepsilon(S,T).$$

This contradicts (3.1.4).

Case 3 $p_0(G-S) = 0$.

Subcase 3.1 $0 \leq p_1(G-S) \leq 2$.

From (3.1.4) and $|S| \geq 2$, we have

$$\varepsilon(S,T) - 1 \geq 2|S| - 2p_0(G-S) - p_1(G-S) \geq 2 \geq \varepsilon(S,T),$$

which is a contradiction.

Subcase 3.2 $p_1(G-S) \geq 3$.

In this case, there must exist $x, y \in P_1(G-S)$ such that $xy \notin E(G)$. According to the hypothesis of Theorem 3.1.3, we obtain

$$\max\{d_G(x), d_G(y)\} \geq \frac{n}{2}.$$

Suppose that $d_G(y) \geq \frac{n}{2}$, we have

$$\frac{n}{2} \leq d_G(y) \leq d_{G-S}(y) + |S| = |S| + 1,$$

which implies $|S| \geq \frac{n}{2} - 1$. Furthermore,

$$p_1(G-S) \leq n - |S| \leq \frac{n}{2} + 1.$$

By (3.1.4) and $n \geq 9$, we obtain

$$1 \geq \varepsilon(S,T) - 1 \geq 2|S| - p_1(G-S)$$

$$\geq 2\left(\frac{n}{2}-1\right)-\left(\frac{n}{2}+1\right)=\frac{n}{2}-3\geq\frac{3}{2}.$$

It is a contradiction. This completes the proof of Theorem 3.1.3. □

Wang[19] obtained a degree condition for the existence of k-deleted graphs.

Theorem 3.1.4 [19]   *Let k be an integer such that $k \geq 3$, and let G be a 2-connected graph of order n with $n \geq 4k + 1$, kn even, and $\delta(G) \geq k + 1$. Suppose that*
$$\max\{d_G(x), d_G(y)\} \geq \frac{n}{2}$$
for each pair of nonadjacent vertices x, y of G. Then G is a k-deleted graph.

Zhou[33] extended k-deleted graphs in Theorem 3.1.4 to fractional k-deleted graphs, and obtained the following the result on fractional k-deleted graphs.

Theorem 3.1.5 [33]   *Let $k \geq 1$ be an integer, and let G be a 2-connected graph of order n with $n \geq \max\{7, 4k + 1\}$, and $\delta(G) \geq k + 1$. If G satisfies*
$$\max\{d_G(x), d_G(y)\} \geq \frac{n}{2}$$
for each pair of nonadjacent vertices x, y of G, then G is a fractional k-deleted graph.

Proof   By Theorem 3.1.2 and Theorem 3.1.3, the result obviously holds for $k = 1$ and $k = 2$. If $k \geq 3$ and kn is even, then G is a k-deleted graph by Theorem 3.1.4. We have known that a k-deleted graph is a special fractional k-deleted graph. Hence, G is a fractional k-deleted graph. Now we consider the case that k and n are both odd.

Suppose that G satisfies the assumption of the theorem, but it is not a fractional k-deleted graph. From Theorem 3.1.1, there exists a subset S of $V(G)$ such that
$$\delta_G(S,T) = k|S| + d_{G-S}(T) - k|T| \leq \varepsilon(S,T) - 1, \tag{3.1.8}$$

3.1 Fractional k-Deleted Graphs

where $T = \{x : x \in V(G) \setminus S, d_{G-S}(x) \leq k\}$. Firstly, we prove the following claims.

Claim 4 $S \neq \emptyset$.

Proof If $S = \emptyset$, then by $\delta(G) \geq k+1$ and (3.1.8) we have

$$\varepsilon(S,T) - 1 \geq \delta_G(S,T) = k|S| + d_{G-S}(T) - k|T|$$
$$= d_G(T) - k|T| \geq (\delta(G) - k)|T|$$
$$\geq |T| \geq \varepsilon(S,T),$$

which is a contradiction. □

Claim 5 $|T| \geq k+1$.

Proof Suppose that $|T| \leq k$. Then by $\delta(G) \geq k+1$, we obtain

$$\delta_G(S,T) = k|S| + d_{G-S}(T) - k|T|$$
$$\geq |T||S| + d_{G-S}(T) - k|T|$$
$$= \sum_{x \in T}(|S| + d_{G-S}(x) - k) \geq \sum_{x \in T}(\delta(G) - k)$$
$$\geq |T| \geq \varepsilon(S,T).$$

This contradicts (3.1.8). □

Claim 6 $|T| \geq |S| + 1$.

Proof If $|T| \leq |S|$, then by (3.1.8) we have

$$\varepsilon(S,T) - 1 \geq \delta_G(S,T) = k|S| + d_{G-S}(T) - k|T|$$
$$\geq d_{G-S}(T) \geq \varepsilon(S,T),$$

which is a contradiction. □

Claim 7 $|S| \leq \dfrac{n-1}{2}$.

Proof According to Claim 6 and $|S| + |T| \leq n$, we obtain

$$n \geq |S| + |T| \geq 2|S| + 1,$$

which implies

$$|S| \leq \frac{n-1}{2}.$$

The proof of Claim 7 is complete. □

According to the hypothesis of Theorem 3.1.5 and since n is odd, we have
$$\max\{d_G(x), d_G(y)\} \geq \frac{n+1}{2} \qquad (3.1.9)$$
for each pair of nonadjacent vertices x, y of G.

By Claim 5, $T \neq \emptyset$. Now we define
$$h_1 = \min\{d_{G-S}(x) : x \in T\},$$
and let x_1 be a vertex in T satisfying $d_{G-S}(x_1) = h_1$. Furthermore, if $T \setminus N_T[x_1] \neq \emptyset$, we define
$$h_2 = \min\{d_{G-S}(x) : x \in T \setminus N_T[x_1]\},$$
and let x_2 be a vertex in $T \setminus N_T[x_1]$ satisfying $d_{G-S}(x_2) = h_2$. Then we have $0 \leq h_1 \leq h_2 \leq k$ by the definition of T and $d_G(x_i) \leq |S| + h_i$ for $i = 1, 2$.

If $T \setminus N_T[x_1] \neq \emptyset$, then we have $|S| + h_2 \geq \frac{n+1}{2}$. Otherwise we get
$$|S| + h_1 \leq |S| + h_2 < \frac{n+1}{2},$$
and this implies $d_G(x_1) < \frac{n+1}{2}$ and $d_G(x_2) < \frac{n+1}{2}$. Since $x_1 x_2 \notin E(G)$, that would contradict (3.1.9).

Now in order to prove the correctness of Theorem 3.1.5, we will deduce some contradictions according to the following cases.

Case 1 $T = N_T[x_1]$.

According to Claim 5, we know that T is not independent. In this case, $\varepsilon(S, T) = 2$. In view of Claim 5 and $T = N_T[x_1]$, we have
$$k \geq h_1 = d_{G-S}(x_1) \geq |T| - 1 \geq k.$$
Hence, $h_1 = k$. By (3.1.8), Claim 4 and the definition of h_1, we obtain
$$1 = \varepsilon(S,T) - 1 \geq \delta_G(S,T) = k|S| + d_{G-S}(T) - k|T|$$
$$\geq k|S| + h_1|T| - k|T| = k|S|$$
$$\geq k \geq 3,$$
a contradiction.

Case 2 $T \setminus N_T[x_1] \neq \emptyset$.

3.1 Fractional k-Deleted Graphs

By Claim 7 and $|S| + h_2 \geq \frac{n+1}{2}$, we have

$$\frac{n-1}{2} \geq |S| \geq \frac{n+1}{2} - h_2,$$

that is,

$$h_2 \geq 1. \qquad (3.1.10)$$

According to (3.1.10), $n \geq 4k+1$, $|S|+|T| \leq n$, $|S|+h_2 \geq \frac{n+1}{2}$, $h_1 \leq h_2 \leq k$ and $|N_T[x_1]| \leq d_{G-S}(x_1)+1 = h_1+1$, we obtain

$$\begin{aligned}
\delta_G(S,T) &= k|S| + d_{G-S}(T) - k|T| \\
&\geq |S| + h_1|N_T[x_1]| + h_2(|T| - |N_T[x_1]|) - k|T| \\
&= k|S| + (h_1 - h_2)|N_T[x_1]| + (h_2 - k)|T| \\
&\geq k|S| + (h_1 - h_2)(h_1 + 1) + (h_2 - k)(n - |S|) \\
&= (2k - h_2)|S| + (h_1 - h_2)(h_1 + 1) + (h_2 - k)n \\
&\geq (2k - h_2)\left(\frac{n+1}{2} - h_2\right) + (h_1 - h_2)(h_1 + 1) + (h_2 - k)n \\
&= \left(\frac{h_2 - 1}{2} - h_1\right)^2 + \frac{3}{4}h_2^2 + \left(\frac{n}{2} - 2k - 1\right)h_2 + k - \frac{1}{4} \\
&\geq \frac{3}{4}h_2^2 + \left(\frac{n}{2} - 2k - 1\right)h_2 + k - \frac{1}{4} \\
&\geq \frac{3}{4}h_2^2 - \frac{1}{2}h_2 + k - \frac{1}{4} \geq \frac{3}{4} - \frac{1}{2} + k - \frac{1}{4} \\
&= k > 2 \geq \varepsilon(S,T),
\end{aligned}$$

this contradicts (3.1.8).

From all the cases above, we deduce the contradictions. Hence, G is a fractional k-deleted graph. This completes the proof of Theorem 3.1.5. □

Remark 3.1.1 Let us show that the condition

$$\max\{d_G(x), d_G(y)\} \geq \frac{n}{2}$$

in Theorem 3.1.5 cannot be replaced by

$$\max\{d_G(x), d_G(y)\} \geq \frac{n}{2} - 1.$$

Let $t \geq 2$ and $k \geq 1$ be two integers. We construct a graph

$$G = ((kt-2)K_1 \cup K_2) \vee (kt+1)K_1.$$

Obviously, G is 2-connected, $\delta(G) = kt \geq 2k \geq k+1$, $n = |V(G)| = 2kt + 1 \geq 4k + 1$ and

$$\frac{n}{2} > \max\{d_G(x), d_G(y)\} > \frac{n}{2} - 1$$

for each pair of nonadjacent vertices x, y of $(kt+1)K_1 \subset G$. Let

$$G' = G - E(K_2), \quad S = V((kt-2)K_1 \cup K_2)) \subseteq V(G)$$

and $T = V((kt+1)K_1) \subseteq V(G)$. Then $|S| = kt$, $|T| = kt + 1$ and $d_{G'-S}(T) = 0$. Thus, we get

$$\delta_{G'}(S, T) = k|S| + d_{G'-S}(T) - k|T|$$
$$= k^2 t - k(kt+1)$$
$$= -k < 0.$$

By Theorem 2.1.1, G' has no fractional k-factor, that is, Theorem 3.1.5 does not hold. In the above sense, the result in Theorem 3.1.5 is best possible.

Remark 3.1.2 In the following, we show that the bound on n in Theorem 3.1.5 is also best possible. Let $k \geq 2$ be an integer. We construct a graph $G = K_{2k-1} \vee (K_1 \cup kK_2)$. Then G satisfies all conditions of Theorem 3.1.5 except $n = 4k$. Let $S = V(K_{2k-1})$ and $T = V(K_1 \cup kK_2)$. Clearly, $d_{G-S}(T) = 2k$ and T is not independent. In this case, $\varepsilon(S,T) = 2$. Thus, we obtain

$$\delta_G(S,T) = k|S| + d_{G-S}(T) - k|T|$$
$$= k(2k-1) + 2k - k(2k+1)$$
$$= 0 < 2 = \varepsilon(S,T).$$

According to Theorem 3.1.1, Theorem 3.1.5 does not hold.

Next we present a neighborhood condition for a graph to be a fractional k-deleted graph, which is the following theorem.

Theorem 3.1.6 [37] *Let k be a positive integer and G a graph of order n with $n \geq 4k - 5$. If*

3.1 Fractional k-Deleted Graphs

$$|N_G(X)| > \frac{(k-1)n + |X| + 1}{2k - 1}$$

for every non-empty independent subset X of $V(G)$, and

$$\delta(G) > \frac{(k-1)(n+2) + 2}{2k - 1},$$

then G is a fractional k-deleted graph.

The proof of Theorem 3.1.6 is quite similar to that of Theorem 2.1.8 and is omitted. In the following, we discuss the relationship between the binding number and fractional k-deleted graphs, and obtain a new result on fractional k-deleted graphs by using the binding number.

Lemma 3.1.1 [21] *Let G be a graph of order n with $\text{bind}(G) > c$. Then*

$$\delta(G) > n - \frac{n-1}{c}.$$

Theorem 3.1.7 [29] *Let $k \geq 2$ be an integer, and let G be a graph of order n with $n \geq 4k - 5$. If*

$$\text{bind}(G) > \frac{(2k-1)(n-1)}{k(n-2)},$$

then G is a fractional k-deleted graph.

Proof Suppose that G satisfies the assumption of the theorem, but it is not a fractional k-deleted graph. Then by Theorem 3.1.1, there exist some $S \subseteq V(G)$ and $T = \{x : x \in V(G) \setminus S, d_{G-S}(x) \leq k\}$ such that

$$\delta_G(S,T) = k|S| + d_{G-S}(T) - k|T| \leq \varepsilon(S,T) - 1. \quad (3.1.11)$$

Claim 1 $|T| \geq k + 1$.

Proof In view of Lemma 3.1.1, we have

$$|S| + d_{G-S}(x) \geq d_G(x) \geq \delta(G) > n - \frac{n-1}{\frac{(2k-1)(n-1)}{k(n-2)}}$$

$$= n - \frac{k(n-2)}{2k-1} = \frac{(k-1)n + 2k}{2k - 1}$$

$$\geq \frac{(k-1)(4k-5)+2k}{2k-1} = 2(k-1) - \frac{k-3}{2k-1}.$$

If $k \geq 3$, then according to the integrity of $\delta(G)$ we obtain

$$|S| + d_{G-S}(x) \geq \delta(G) \geq 2k - 2. \tag{3.1.12}$$

If $k = 2$, then by the integrity of $\delta(G)$ we get

$$|S| + d_{G-S}(x) \geq \delta(G) \geq 2k - 1. \tag{3.1.13}$$

Let $|T| \leq k$ and $k \geq 3$, then by (3.1.11) and (3.1.12), we have

$$\varepsilon(S,T) - 1 \geq \delta_G(S,T) = k|S| + d_{G-S}(T) - k|T|$$

$$\geq |T||S| + d_{G-S}(T) - k|T|$$

$$= \sum_{x \in T}(|S| + d_{G-S}(x) - k) \geq \sum_{x \in T}(2k - 2 - k)$$

$$= \sum_{x \in T}(k-2) = (k-2)|T| \geq |T| \geq \varepsilon(S,T),$$

which is a contradiction.

Let $|T| \leq k$ and $k = 2$, then by (3.1.11) and (3.1.13), we have

$$\varepsilon(S,T) - 1 \geq \delta_G(S,T) = k|S| + d_{G-S}(T) - k|T|$$

$$\geq |T||S| + d_{G-S}(T) - k|T|$$

$$= \sum_{x \in T}(|S| + d_{G-S}(x) - k) \geq \sum_{x \in T}(2k - 1 - k)$$

$$= \sum_{x \in T}(k-1) = (k-1)|T| = |T| \geq \varepsilon(S,T),$$

a contradiction. □

Claim 2 $S \neq \emptyset$.

Proof Let $S = \emptyset$. If $k \geq 3$, then by (3.1.11) and (3.1.12) we get that

$$\varepsilon(S,T) - 1 \geq \delta_G(S,T) = k|S| + d_{G-S}(T) - k|T|$$

$$= d_G(T) - k|T| \geq (\delta(G) - k)|T|$$

$$\geq (2k - 2 - k)|T| = (k-2)|T|$$

$$\geq |T| \geq \varepsilon(S,T),$$

this is a contradiction.

If $k = 2$, then by (3.1.11) and (3.1.13) we have

$$\varepsilon(S,T) - 1 \geq \delta_G(S,T) = k|S| + d_{G-S}(T) - k|T|$$
$$= d_G(T) - k|T| \geq (\delta(G) - k)|T|$$
$$\geq (2k - 1 - k)|T| = (k-1)|T|$$
$$= |T| \geq \varepsilon(S,T).$$

Which is a contradiction. □

Claim 3 There exists $x \in T$ such that $d_{G-S}(x) \leq k-1$.

Proof If $d_{G-S}(x) \geq k$ for all $x \in T$, then we get from Claim 2

$$\delta_G(S,T) = k|S| + d_{G-S}(T) - k|T| \geq k|S| \geq k \geq 2 \geq \varepsilon(S,T),$$

which contradicts (3.1.11). □

Define $h = \min\{d_{G-S}(x) : x \in T\}$. Then by Claim 3, we obtain

$$0 \leq h \leq k-1.$$

By Lemma 3.1.1 and $\delta(G) \leq |S| + h$, we get

$$|S| \geq \delta(G) - h > n - \frac{k(n-2)}{2k-1} - h = \frac{(k-1)n + 2k}{2k-1} - h. \quad (3.1.14)$$

The proof splits into two cases.

Case 1 $h = 0$.

First, we prove the following claim.

Claim 4 $\dfrac{k(n-2)}{n-1} \geq 1$.

Proof In view of $k \geq 2$ and $n \geq 4k - 5$, we get

$$k(n-2) - (n-1) = (k-1)(n-2) - 1 \geq 0.$$

Thus, we obtain

$$\frac{k(n-2)}{n-1} \geq 1.$$

This completes the proof of Claim 4. □

Let m be the number of vertices x in T such that $d_{G-S}(x) = 0$, and let $Y = V(G) \setminus S$. Then $N_G(Y) \neq V(G)$ since $h = 0$, and $Y \neq \emptyset$ by Claim 1, and so $|N_G(Y)| \geq \text{bind}(G)|Y|$. Thus

$$n - m \geq |N_G(Y)| \geq \text{bind}(G)|Y| = \text{bind}(G)(n - |S|),$$

that is,

$$|S| \geq n - \frac{n-m}{\text{bind}(G)} > n - \frac{k(n-2)(n-m)}{(2k-1)(n-1)}. \qquad (3.1.15)$$

According to (3.1.11), (3.1.15), Claim 4 and $|T| \leq n - |S|$, we have

$$\varepsilon(S, T) - 1 \geq \delta_G(S, T) = k|S| + d_{G-S}(T) - k|T|$$
$$\geq k|S| - k|T| + |T| - m$$
$$\geq k|S| - (k-1)(n - |S|) - m$$
$$= (2k-1)|S| - kn + n - m$$
$$> (2k-1)\left(n - \frac{k(n-2)(n-m)}{(2k-1)(n-1)}\right) - kn + n - m$$
$$= kn - \frac{k(n-2)(n-m)}{n-1} - m$$
$$\geq kn - \frac{k(n-2)(n-1)}{n-1} - 1$$
$$= kn - k(n-2) - 1 = 2k - 1$$
$$> 2 \geq \varepsilon(S, T).$$

This is a contradiction.

Case 2 $1 \leq h \leq k - 1$.

In view of Claim 1, we obtain

$$|T| \geq k + 1 > h + 1.$$

Let v be a vertex in T such that $d_{G-S}(v) = h$, and put $Y = T - N_{G-S}(v)$. Then $|Y| \geq |T| - h > 1$ and $N_G(Y) \neq V(G)$. Thus, we get

$$\frac{n-1}{|T|-h} \geq \frac{|N_G(Y)|}{|Y|} \geq \text{bind}(G) > \frac{(2k-1)(n-1)}{k(n-2)},$$

that is,

$$|T| < \frac{k(n-2)}{2k-1} + h. \qquad (3.1.16)$$

3.1 Fractional k-Deleted Graphs

By(3.1.14) and (3.1.16), we have

$$\delta_G(S,T) = k|S| + d_{G-S}(T) - k|T| \geq k|S| - k|T| + h|T|$$
$$= k|S| - (k-h)|T|$$
$$> k\left(\frac{(k-1)n+2k}{2k-1} - h\right) - (k-h)\left(\frac{k(n-2)}{2k-1} + h\right).$$

Subcase 2.1 $h = 1$.

Obviously, we obtain

$$\delta_G(S,T) > k\left(\frac{(k-1)n+2k}{2k-1} - 1\right) - (k-1)\left(\frac{k(n-2)}{2k-1} + 1\right)$$
$$= k \cdot \frac{(k-1)n+1}{2k-1} - (k-1) \cdot \frac{kn-1}{2k-1} = \frac{2k-1}{2k-1} = 1.$$

According to the integrity of $\delta_G(S,T)$, we get that

$$\delta_G(S,T) \geq 2 \geq \varepsilon(S,T),$$

this contradicts (3.1.11).

Subcase 2.2 $2 \leq h \leq k-1$.

Clearly, $k \geq 3$. Let

$$f(h) = k\left(\frac{(k-1)n+2k}{2k-1} - h\right) - (k-h)\left(\frac{k(n-2)}{2k-1} + h\right).$$

Then

$$\delta_G(S,T) > f(h), \qquad (3.1.17)$$

and

$$f'(h) = -2k + 2h + \frac{k(n-2)}{2k-1}.$$

Since $2 \leq h \leq k-1$ and $n \geq 4k-5$, we have

$$f'(h) \geq -2k + 4 + \frac{k(n-2)}{2k-1} = \frac{-4k^2 + 2k + 8k - 4 + kn - 2k}{2k-1}$$
$$= \frac{kn - 4k^2 + 8k - 4}{2k-1} \geq \frac{k(4k-5) - 4k^2 + 8k - 4}{2k-1}$$
$$= \frac{3k-4}{2k-1} > 0.$$

Thus, we get
$$f(h) \geq f(2). \qquad (3.1.18)$$

From (3.1.17), (3.1.18) and $k \geq 3$, we obtain

$$\delta_G(S,T) > f(h) \geq f(2)$$
$$= k\left(\frac{(k-1)n+2k}{2k-1} - 2\right) - (k-2)\left(\frac{k(n-2)}{2k-1} + 2\right)$$
$$= \frac{k(k-1)n + 2k^2 - 4k^2 + 2k - k(k-2)n - 2k^2 + 6k - 4}{2k-1}$$
$$= \frac{kn - 4k^2 + 8k - 4}{2k-1} \geq \frac{k(4k-5) - 4k^2 + 8k - 4}{2k-1}$$
$$= \frac{3k-4}{2k-1} = 1 + \frac{k-3}{2k-1} \geq 1.$$

By the integrity of $\delta_G(S,T)$, we have
$$\delta_G(S,T) \geq 2 \geq \varepsilon(S,T),$$
which contradicts (3.1.11).

From all the cases above, we deduced the contradiction. Hence, G is a fractional k-deleted graph. Completing the proof of Theorem 3.1.7. □

Remark 3.1.3 Let us show that the condition
$$\text{bind}(G) > \frac{(2k-1)(n-1)}{k(n-2)}$$
in Theorem 3.1.7 can not be replaced by
$$\text{bind}(G) \geq \frac{(2k-1)(n-1)}{k(n-2)}.$$

Let $r \geq 1$, $k \geq 3$ be two odd positive integer and let $l = \frac{5kr+1}{2}$ and $m = 5kr - 5r + 1$, so that
$$n = m + 2l = 10kr - 5r + 2.$$

Let $H = K_m \bigvee lK_2$ and $X = V(lK_2)$. Then for any $x \in X$,
$$|N_H(X \setminus x)| = n - 1.$$

By the definition of $\text{bind}(H)$,

84

3.1 Fractional k-Deleted Graphs

$$\text{bind}(H) = \frac{|N_H(X \setminus x)|}{|X \setminus x|} = \frac{n-1}{2l-1} = \frac{n-1}{5kr} = \frac{(2k-1)(n-1)}{k(n-2)}.$$

Let $S = V(K_m) \subseteq V(H)$, $T = V(lK_2) \subseteq V(H)$. Then $|S| = m$, $|T| = 2l$. Obviously, T is not independent, then $\varepsilon(S,T) = 2$. Thus, we obtain

$$\begin{aligned}\delta_H(S,T) &= k|S| - k|T| + d_{H-S}(T) = k|S| - k|T| + |T| \\ &= k|S| - (k-1)|T| = km - 2(k-1)l \\ &= k(5kr - 5r + 1) - (k-1)(5kr + 1) \\ &= 1 < 2 = \varepsilon(S,T).\end{aligned}$$

By Theorem 3.1.1, H is not a fractional k-deleted graph. In the above sense, the result in Theorem 3.1.7 is best possible.

Remark 3.1.4 We don't know whether the result can be strengthened to the form that if $\text{bind}(G) > \frac{(2k-1)(n-1)}{k(n-2)}$ then G is k-deleted. We guess that the above result can hold for kn even.

If G satisfies $\delta(G) \geq 1 + \frac{(k-1)n+2}{2k-1}$ in Theorem 3.1.7, then the binding number condition in Theorem 3.1.7 can be improved. Our result is the following theorem.

Theorem 3.1.8 [31] Let $k \geq 2$ be an integer, and let G be a graph of order n with $n \geq 4k - 3$. If G satisfies

$$\text{bind}(G) > 1 + \frac{k-1}{k}$$

and

$$\delta(G) \geq 1 + \frac{(k-1)n+2}{2k-1},$$

then G is a fractional k-deleted graph.

Proof Suppose that G satisfies the assumption of Theorem 3.1.8, but it is not a fractional k-deleted graph. Then by Theorem 3.1.1, there exists a subset S of $V(G)$ such that

$$\delta_G(S,T) = k|S| + d_{G-S}(T) - k|T| \leq \varepsilon(S,T) - 1 \leq 1, \quad (3.1.19)$$

where $T = \{x : x \in V(G) \setminus S, d_{G-S}(x) \leq k\}$.

We first show that the following claims hold.

Claim 1 $S \neq \emptyset$.

Proof Suppose $S = \emptyset$. Then by (3.1.19),

$$\delta(G) \geq 1 + \frac{(k-1)n+2}{2k-1}$$

and $n \geq 4k - 3$, we have

$$\varepsilon(S,T) - 1 \geq \delta_G(S,T) = k|S| + d_{G-S}(T) - k|T|$$
$$= d_G(T) - k|T| \geq (\delta(G) - k)|T|$$
$$\geq \left(1 + \frac{(k-1)n+2}{2k-1} - k\right)|T|$$
$$= \frac{(k-1)(n-2k+1)+2}{2k-1}|T|$$
$$\geq \frac{(k-1)(4k-3-2k+1)+2}{2k-1}|T|$$
$$= \frac{(k-1)(2k-2)+2}{2k-1}|T|$$
$$\geq |T| \geq \varepsilon(S,T),$$

which is a contradiction. □

Claim 2 $T \neq \emptyset$.

Proof If $T = \emptyset$, then by (3.1.19) and Claim 1 we obtain

$$\varepsilon(S,T) - 1 \geq \delta_G(S,T) = k|S| \geq k \geq 2 \geq \varepsilon(S,T).$$

This is a contradiction. □

Define

$$h = \min\{d_{G-S}(x) : x \in T\}.$$

According to the definition of T, we have $0 \leq h \leq k$.

In the following, we shall consider two cases according to the value of h and derive a contradiction in each case.

Case 1 $1 \leq h \leq k$.

In view of (3.1.19) and $|S| + |T| \leq n$, we obtain

$$1 \geq \delta_G(S,T) = k|S| + d_{G-S}(T) - k|T|$$

3.1 Fractional k-Deleted Graphs

$$\geq k|S| + h|T| - k|T| = k|S| - (k-h)|T|$$
$$\geq k|S| - (k-h)(n-|S|)$$
$$= (2k-h)|S| - (k-h)n,$$

which implies,
$$|S| \leq \frac{(k-h)n + 1}{2k - h}. \qquad (3.1.20)$$

From (3.1.20), and $\delta(G) \leq h + |S|$, we get
$$\delta(G) \leq h + |S| \leq h + \frac{(k-h)n + 1}{2k - h}. \qquad (3.1.21)$$

Let
$$f(h) = h + \frac{(k-h)n + 1}{2k - h}.$$

Then by $n \geq 4k - 3$, $1 \leq h \leq k$ and $k \geq 2$, we have
$$f'(h) = 1 + \frac{-(2k-h)n + ((k-h)n + 1)}{(2k-h)^2} = 1 - \frac{kn - 1}{(2k-h)^2}$$
$$\leq 1 - \frac{kn - 1}{(2k-1)^2} \leq 1 - \frac{k(4k-3) - 1}{(2k-1)^2}$$
$$= -\frac{k-2}{(2k-1)^2} \leq 0.$$

Thus, we have
$$f(h) \leq f(1) = 1 + \frac{(k-1)n + 1}{2k - 1}. \qquad (3.1.22)$$

According to (3.1.21), (3.1.22) and the hypothesis of Theorem 3.1.8, we get
$$1 + \frac{(k-1)n + 2}{2k - 1} \leq \delta(G) \leq 1 + \frac{(k-1)n + 1}{2k - 1}.$$

That is a contradiction.

Case 2 $h = 0$.

Let $m = |\{x : x \in T, d_{G-S}(x) = 0\}|$, and let $Y = V(G) \setminus S$. Then $N_G(Y) \neq V(G)$ since $h = 0$. In view of the definition of the binding number $\text{bind}(G)$ and the hypothesis of Theorem 3.1.8, we have
$$|N_G(Y)| \geq \text{bind}(G)|Y| > \left(1 + \frac{k-1}{k}\right)|Y|. \qquad (3.1.23)$$

It is easy to see that $|N_G(Y)| \leq n - m$ and $|Y| = n - |S|$. Complying these with (3.1.23), we obtain

$$n - m \geq |N_G(Y)| > \left(1 + \frac{k-1}{k}\right)|Y| = \left(1 + \frac{k-1}{k}\right)(n - |S|),$$

that is,

$$|S| > n - \frac{k(n-m)}{2k-1}. \tag{3.1.24}$$

From (3.1.24) and $|S| + |T| \leq n$, we have

$$\delta_G(S,T) = k|S| + d_{G-S}(T) - k|T| \geq k|S| + |T| - m - k|T|$$
$$= k|S| - (k-1)|T| - m \geq k|S| - (k-1)(n - |S|) - m$$
$$= (2k-1)|S| - (k-1)n - m$$
$$> (2k-1)\left(n - \frac{k(n-m)}{2k-1}\right) - (k-1)n - m$$
$$= kn - k(n-m) - m \geq kn - k(n-1) - 1$$
$$= k - 1 \geq 1,$$

which contradicts (3.1.19).

From all the cases above, we deduce the contradictions. Hence, G is a fractional k-deleted. This completes the proof of the theorem. $\square$

3.2 Fractional (g, f)-Deleted Graphs

Let G be a graph, and let g, f two integer-valued functions defined on $V(G)$ satisfying $0 \leq g(x) \leq f(x)$ for each $x \in V(G)$. Then a spanning subgraph F of G is called a (g, f)-**factor** if $g(x) \leq d_F(x) \leq f(x)$ for all $x \in V(G)$. For $x \in V(G)$, $E(x)$ denotes the set of edges incident with x. A fractional (g, f)-factor is a function h that assigns to each edge of a graph G a number in [0,1], so that for each vertex x we have

$$g(x) \leq d_G^h(x) \leq f(x),$$

where

3.2 Fractional (g, f)-Deleted Graphs

$$d_G^h(x) = \sum_{e \in E(x)} h(e)$$

is the fractional degree of x in G. A graph G is called a **fractional (g, f)-deleted graph** if $G - e$ has a fractional (g, f)-factor for any $e \in E(G)$. If $g(x) = f(x)$ for each $x \in V(G)$, then a fractional (g, f)-deleted graph is a fractional f-deleted graph. If $g(x) = f(x) = k$ for each $x \in V(G)$, then a fractional (g, f)-deleted graph is a fractional k-deleted graph.

A necessary and sufficient condition for a graph to be a fractional (g, f)-deleted graph was obtained by Li, Yan and Zhang[11] in 2003. In this section, we discuss the existence of fractional (g, f)-deleted graphs, which is a generalization of fractional k-deleted graphs. We present some sufficient conditions for graphs to be fractional (g, f)-deleted graphs and show the results are best possible in some sense.

Theorem 3.2.1 [11] *Let G be a graph, and let g and f be two nonnegative integer-valued functions defined on $V(G)$ such that $g(x) \leq f(x)$ for any $x \in V(G)$. Then G is a fractional (g, f)-deleted graph if and only if*

$$\delta_G(S, T) = f(S) + d_{G-S}(T) - g(T) \geq \varepsilon(S, T)$$

for any $S \subseteq V(G)$, where $T = \{x \in V(G) - S : d_{G-S}(x) \leq g(x)\}$, and $\varepsilon(S, T)$ is defined as follows,

$$\varepsilon(S, T) = \begin{cases} 2, & \text{if } T \text{ is not independent,} \\ 1, & \text{if } T \text{ is independent, and } e_G(T, V(G) - (S \cup T)) \geq 1, \\ 0, & \text{otherwise.} \end{cases}$$

By using Theorem 3.2.1, Zhou[32, 39] verified three results on the existence of fractional (g, f)-deleted graphs, which are shown in the following.

We first present two sufficient conditions related to binding number for graphs to be fractional (g, f)-deleted graphs.

Theorem 3.2.2 [32] *Let a and b be two integers with $2 \leq a \leq b$, and let g and f be two integer-valued functions defined on $V(G)$ such that $a \leq g(x) \leq f(x) \leq b$ for any $x \in V(G)$. Let G be a graph of order n with*

$n \geq \frac{(a+b-2)^2+2}{a}$, and suppose that
$$\text{bind}(G) > \frac{(a+b-1)(n-1)}{an-(a+b)}.$$
Then G is a fractional (g,f)-deleted graph.

Moreover, in the case where a and b are even, we can relax the binding number condition as follows:

Theorem 3.2.3 [32] Let a and b be two even integers with $2 \leq a \leq b$, and let g and f be two integer-valued functions defined on $V(G)$ such that $a \leq g(x) \leq f(x) \leq b$ for any $x \in V(G)$. Let G be a graph of order n with $n \geq \frac{(a+b-1)(a+b-3)+2}{a}$, and suppose that
$$\text{bind}(G) > \frac{(a+b-1)(n-1)}{an-(a+b)+1}.$$
Then G is a fractional (g,f)-deleted graph.

If $g(x) = f(x)$ for any $x \in V(G)$ in Theorem 3.2.2 and Theorem 3.2.3, then we get immediately the following corollary.

Corollary 4 Let a and b be two integers with $2 \leq a \leq b$, and let f be an integer-valued functions defined on $V(G)$ such that $a \leq f(x) \leq b$ for any $x \in V(G)$. Let G be a graph of order n with $n \geq \frac{(a+b-2)^2+2}{a}$, and suppose that
$$\text{bind}(G) > \frac{(a+b-1)(n-1)}{an-(a+b)}.$$
Then G is a fractional f-deleted graph.

Corollary 5 Let a and b be two even integers with $2 \leq a \leq b$, and let f be an integer-valued functions defined on $V(G)$ such that $a \leq f(x) \leq b$ for any $x \in V(G)$. Let G be a graph of order n with $n \geq \frac{(a+b-1)(a+b-3)+2}{a}$, and suppose that
$$\text{bind}(G) > \frac{(a+b-1)(n-1)}{an-(a+b)+1}.$$
Then G is a fractional f-deleted graph.

3.2 Fractional (g, f)-Deleted Graphs

The proof of Theorem 3.2.2 is quite similar to that of Theorem 3.2.3. Hence, we prove only Theorem 3.2.3 and use the similar proof method in [9].

Proof of Theorem 3.2.3 Suppose that G is not a fractional (g, f)-deleted graph. Then by Theorem 3.2.1, there exists some subset $S \subseteq V(G)$ such that

$$\delta_G(S, T) = f(S) + d_{G-S}(T) - g(T) \leq \varepsilon(S, T) - 1, \qquad (3.2.1)$$

where $T = \{x : x \in V(G) - S, d_{G-S}(x) \leq g(x)\}$.

In terms of the hypotheses of Theorem 3.2.3 and Lemma 3.1.1, we obtain

$$\delta(G) > n - \frac{an - (a+b) + 1}{a+b-1}$$

$$= \frac{(b-1)n + (a+b) - 1}{a+b-1} \geq b. \qquad (3.2.2)$$

Claim 1 $T \neq \emptyset$.

Proof If $T = \emptyset$, then $\varepsilon(S, T) = 0$. According to (3.2.1), we have

$$-1 \geq \delta_G(S, T) = f(S) \geq 0.$$

It is a contradiction. This completes the proof of Claim 1. □

Claim 2 $S \neq \emptyset$.

Proof Assume that $S = \emptyset$. Then by (3.2.2), we get

$$\delta_G(S, T) = d_G(T) - g(T) \geq (b+1)|T| - b|T| = |T| \geq \varepsilon(S, T),$$

which contradicts (3.2.1). This completes the proof of Claim 2. □

Since $T \neq \emptyset$, we define

$$h = \min\{d_{G-S}(t) : t \in T\}.$$

From the definition of T, we have $0 \leq h \leq b$.

In the following, we shall consider various cases by the value of h and find out a contradiction in every case.

Case 1 $h = 0$.

We set $Y = \{t : t \in T, d_{G-S}(t) = 0\}$, then $Y \neq \emptyset$ and $N_G(V(G) - S) \cap Y = \emptyset$. Hence

$$\text{bind}(G) \leq \frac{|N_G(V(G) - S)|}{|V(G) - S|} \leq \frac{n - |Y|}{n - |S|}. \qquad (3.2.3)$$

Using Claim 2, we have $|Y| < n$. Combining this with (3.2.3), we get

$$|S| \geq n - \frac{n-|Y|}{\text{bind}(G)} > n - \frac{(n-|Y|)(an-(a+b)+1)}{(a+b-1)(n-1)}$$

$$> \frac{(b-1)n + a|Y|}{a+b-1}. \tag{3.2.4}$$

According to (3.2.4), $|S| + |T| \leq n$ and $Y \neq \emptyset$, we obtain

$$\delta_G(S,T) = f(S) + d_{G-S}(T) - g(T) \geq a|S| + |T| - |Y| - b|T|$$

$$= a|S| - (b-1)|T| - |Y| \geq a|S| - (b-1)(n-|S|) - |Y|$$

$$= (a+b-1)|S| - (b-1)n - |Y|$$

$$> (a+b-1) \cdot \frac{(b-1)n + a|Y|}{a+b-1} - (b-1)n - |Y|$$

$$= (a-1)|Y| \geq a - 1 \geq 1.$$

In terms of the integrity of $\delta_G(S,T)$ and the definition of $\varepsilon(S,T)$, we have

$$\delta_G(S,T) \geq 2 \geq \varepsilon(S,T),$$

that contradicts (3.2.1).

Case 2 $1 \leq h \leq b-1$.

In this case, we first prove the following claim.

Claim 3 $|S| \leq \frac{(b-h)n}{a+b-h}$.

Proof From (3.2.1) and $\varepsilon(S,T) \leq 2$, we have

$$1 \geq \delta_G(S,T) = f(S) + d_{G-S}(T) - g(T) \geq a|S| + h|T| - b|T|,$$

and hence

$$a|S| - (b-h)|T| \leq 1. \tag{3.2.5}$$

Assume that h is even. Since a and b are even, the LHS of (3.2.5) is even. Thus, we obtain

$$a|S| - (b-h)|T| \leq 0. \tag{3.2.6}$$

In terms of (3.2.6) and $|S| + |T| \leq n$, we get

3.2 Fractional (g, f)-Deleted Graphs

$$|S| \leq \frac{(b-h)n}{a+b-h}.$$

In the following, we assume that h is odd.

If there exists $x \in T$ such that $d_{G-S}(x) \geq h+1$, then we have

$$d_{G-S}(T) \geq h|T| + 1. \tag{3.2.7}$$

According to (3.2.1), (3.2.7), $|S| + |T| \leq n$ and $\varepsilon(S, T) \leq 2$, we obtain

$$1 \geq \delta_G(S, T) = f(S) + d_{G-S}(T) - g(T)$$
$$\geq a|S| + h|T| + 1 - b|T| = a|S| - (b-h)|T| + 1$$
$$\geq a|S| - (b-h)(n - |S|) + 1$$
$$= (a + b - h)|S| - (b-h)n + 1,$$

which implies

$$|S| \leq \frac{(b-h)n}{a+b-h}.$$

If $V(G) - S - T \neq \emptyset$, then we get

$$|S| + |T| \leq n - 1. \tag{3.2.8}$$

Using (3.2.5), (3.2.8) and $1 \leq h \leq b - 1$, we obtain

$$1 \geq a|S| - (b-h)|T| \geq a|S| - (b-h)(n - 1 - |S|)$$
$$= (a + b - h)|S| - (b-h)n + (b-h)$$
$$\geq (a + b - h)|S| - (b-h)n + 1,$$

that is,

$$|S| \leq \frac{(b-h)n}{a+b-h}.$$

If $V(G) - S - T = \emptyset$ and $d_{G-S}(x) = h$ for each $x \in T$, then $d_{G[T]}(x) = h$ for each $x \in T$. Since h is odd, $|T|$ is even. Thus, the LHS of (3.2.5) is even. Hence, we have

$$a|S| - (b-h)|T| \leq 0.$$

Combining this with $|S| + |T| = n$, we get

$$|S| \leq \frac{(b-h)n}{a+b-h}.$$

This completes the proof of Claim 3. □

According to $\delta(G) \leq |S| + h$ and (3.2.2), we obtain

$$|S| \geq \delta(G) - h > \frac{(b-1)n + (a+b) - 1}{a+b-1} - h. \qquad (3.2.9)$$

Using (3.2.9) and Claim 3, we obtain

$$\frac{(b-1)n + (a+b)}{a+b-1} - h \leq |S| \leq \frac{(b-h)n}{a+b-h}. \qquad (3.2.10)$$

If the LHS and RHS of (3.2.10) are denoted by A and B respectively, then (3.2.10) says that $A - B \leq 0$. But, after some rearranging, we find that

$$(a+b-1)(a+b-h)(A-B)$$
$$= (h-1)(an - (a+b-1)(a+b-h) - 1)$$
$$+ a + b - 1. \qquad (3.2.11)$$

Clearly this is positive if $h = 1$. Recall that

$$n \geq \frac{(a+b-1)(a+b-3) + 2}{a}$$

by a hypothesis of the theorem. If $h = 2$, then

$$(a+b-1)(a+b-h)(A-B)$$
$$= an - (a+b-1)(a+b-2) - 1 + a + b - 1$$
$$\geq 1 > 0.$$

If $h \geq 3$, then

$$an - (a+b-1)(a+b-h) - 1$$
$$\geq an - (a+b-1)(a+b-3) - 1 \geq 1 > 0.$$

Hence,

$$(a+b-1)(a+b-h)(A-B) \geq 2 + a + b - 1 > 0.$$

Obviously, the expression in (3.2.11) is positive in all cases, and this contradicts (3.2.10).

Case 3 $h = b$.

According to (3.2.1), $\varepsilon(S,T) \leq 2$ and Claim 2, we obtain

$$1 \geq \varepsilon(S,T) - 1 \geq \delta_G(S,T) \geq a|S| + h|T| - b|T| = a|S| \geq a \geq 2,$$

that is a contradiction.

From all the cases above, we deduce the contradictions. Hence, G is a fractional (g, f)-deleted graph. This completes the proof of Theorem 3.2.3. $\square$

Remark 3.2.1 We now show that the condition
$$\text{bind}(G) > \frac{(a+b-1)(n-1)}{an-(a+b)}$$
in Corollary 4 is best possible. Let a, b and r be three odd integers with $3 \le a \le b$ and $b = ar$, and let m is a sufficiently larger odd integer. Let
$$G = K_{mb-m+r} \vee \frac{ma+1}{2} K_2.$$
We write n for the order of G. Then $n = m(a+b-1)+r+1$. Set
$$X = V\left(\frac{ma+1}{2} K_2\right).$$
Then for any $x \in X$, we have $|N_G(X \setminus x)| = n - 1$. In terms of the definition of $\text{bind}(G)$ and $n = m(a+b-1)+r+1$,
$$\text{bind}(G) = \frac{|N_G(X \setminus x)|}{|X \setminus x|} = \frac{n-1}{ma} = \frac{(a+b-1)(n-1)}{ma(a+b-1)}$$
$$= \frac{(a+b-1)(n-1)}{an-(a+b)}.$$
Let $S = V(K_{mb-m+r})$ and $T = V\left(\frac{ma+1}{2} K_2\right)$. Then
$$|S| = m(b-1)+r, \quad |T| = ma+1$$
and $\varepsilon(S,T) = 2$. Define a function $f : V(G) \to \{a, a+1, \cdots, b\}$ by
$$f(x) = \begin{cases} a, & \text{if } x \in V(K_{mb-m+r}), \\ b, & \text{otherwise.} \end{cases}$$
Thus, we have
$$\delta_G(S,T) = f(S) + d_{G-S}(T) - f(T)$$
$$= a|S| + |T| - b|T| = a|S| - (b-1)|T|$$
$$= a(m(b-1)+r) - (b-1)(ma+1)$$
$$= 1 < 2 = \varepsilon(S,T).$$

Using Theorem 3.2.1, G is not a fractional f-deleted graph. In the sense above, the result of Corollary 4 is best possible. And so the result of Theorem 3.2.2 is also best possible.

Remark 3.2.2 We now show that the condition

$$\text{bind}(G) > \frac{(a+b-1)(n-1)}{an-(a+b)+1}$$

in Corollary 5 is best possible. Let a and b be two even integers with $2 \leq a \leq b$, and let G be a graph of order n. Let

$$G = K_{mb-m} \bigvee \frac{ma}{2} K_2,$$

where m is a sufficiently larger integer. Set $X = V\left(\frac{ma}{2} K_2\right)$. Note that for any $x \in X$, we have $|N_G(X \setminus x)| = n - 1$. From the definition of $\text{bind}(G)$ and $n = m(a + b - 1)$,

$$\text{bind}(G) = \frac{|N_G(X \setminus x)|}{|X \setminus x|} = \frac{n-1}{ma-1} = \frac{(a+b-1)(n-1)}{(a+b-1)(ma-1)}$$

$$= \frac{(a+b-1)(n-1)}{an-(a+b)+1}.$$

Let $S = V(K_{mb-m})$ and $T = V\left(\frac{ma}{2} K_2\right)$. Then $|S| = m(b-1)$, $|T| = ma$ and $\varepsilon(S,T) = 2$. Define a function $f : V(G) \to \{a, a+1, \cdots, b\}$ by

$$f(x) = \begin{cases} a, & \text{if } x \in V(K_{mb-m}), \\ b, & \text{otherwise}. \end{cases}$$

Thus, we obtain

$$\delta_G(S,T) = f(S) + d_{G-S}(T) - f(T)$$
$$= a|S| + |T| - b|T| = a|S| - (b-1)|T|$$
$$= a \cdot m(b-1) - (b-1) \cdot ma$$
$$= 0 < 2 = \varepsilon(S,T).$$

According to Theorem 3.2.1, G is not a fractional f-deleted graph. In the sense above, the result of Corollary 5 is best possible. And so the result of Theorem 3.2.3 is also best possible.

In the following, we present a sufficient condition for a graph to be a

3.2 Fractional (g, f)-Deleted Graphs

fractional (g, f)-deleted graph, which was obtained by Zhou and Bian[39].

Theorem 3.2.4 [39] Let $r \geq 2$, $\beta \geq 0$ and $2 \leq a \leq b - \beta$ be integers, and let G be a graph of order n with

$$n \geq \frac{(a+b)(a+b+1+(r-2)(b-\beta))}{a+\beta}$$

and

$$\delta(G) \geq \frac{(r-1)(b-\beta)(b+1)}{a+\beta},$$

and let g and f be two integer-valued functions defined on $V(G)$ satisfying $a \leq g(x) \leq f(x) - \beta \leq b - \beta$ for each $x \in V(G)$. If G satisfies

$$\max\{d_G(x_1), d_G(x_2), \cdots, d_G(x_r)\} \geq \frac{(b-\beta)n}{a+b}$$

for any independent subset $\{x_1, x_2, \cdots, x_r\}$ in G, then G is a fractional (g, f)-deleted graph.

Proof Suppose that G satisfies the assumption of Theorem 3.2.4, but it is not a fractional (g, f)-deleted graph. Then by Theorem 3.1.1, there exists some subset $S \subseteq V(G)$ such that

$$\delta_G(S, T) = f(S) + d_{G-S}(T) - g(T) \leq \varepsilon(S, T) - 1, \quad (3.2.12)$$

where $T = \{x : x \in V(G) \setminus S, d_{G-S}(x) \leq g(x)\}$.

Claim 1 $T \neq \emptyset$.

Proof If $T = \emptyset$, then $\varepsilon(S, T) = 0$ by the definition of $\varepsilon(S, T)$. Combining this with (3.2.12), we obtain

$$-1 = \varepsilon(S, T) - 1 \geq \delta_G(S, T) = f(S) \geq 0,$$

which is a contradiction. This completes the proof of Claim 1. □

Since $T \neq \emptyset$ (by Claim 1), we may construct a sequence $x_1, x_2, \cdots, x_\lambda$ of vertices of T. Define

$$h_1 = \min\{d_{G-S}(x) : x \in T\}$$

and choose $x_1 \in T$ with $d_{G-S}(x_1) = h_1$. If $\lambda \geq 2$ and $T \setminus (\bigcup_{i=1}^{\lambda-1} N_T[x_i]) \neq \emptyset$, define

$$h_\lambda = \min\left\{d_{G-S}(x) : x \in T \setminus \left(\bigcup_{i=1}^{\lambda-1} N_T[x_i]\right)\right\}$$

and choose $x_\lambda \in T \setminus (\bigcup_{i=1}^{\lambda-1} N_T[x_i])$ with $d_{G-S}(x_\lambda) = h_\lambda$, $2 \le \lambda \le r$. Note that $0 \le h_1 \le h_2 \le \cdots \le h_\lambda \le b - \beta$ hold and the subset $\{x_1, x_2, \cdots, x_\lambda\}$ of T is independent.

Claim 2 $|T| \ge \begin{cases} (r-1)(b+1), & \text{if } d_{G-S}(x) = 1 \text{ for any } x \in T, \\ (r-1)(b+1) + 1, & \text{otherwise.} \end{cases}$

Proof Note that

$$|S| + h_1 = |S| + d_{G-S}(x_1) \ge d_G(x_1) \ge \delta(G)$$
$$\ge \frac{(r-1)(b-\beta)(b+1)}{a+\beta},$$

which implies

$$|S| \ge \frac{(r-1)(b-\beta)(b+1)}{a+\beta} - h_1. \qquad (3.2.13)$$

We first prove that

$$|T| \ge (r-1)(b+1)$$

if $d_{G-S}(x) = 1$ for any $x \in T$. In this case, $h_1 = 1$.

Assume that $|T| \le (r-1)(b+1) - 1$. According to (3.2.12), (3.2.13), $h_1 = 1$ and $r \ge 2$, we have

$$\varepsilon(S,T) - 1 \ge \delta_G(S,T) = f(S) + d_{G-S}(T) - g(T)$$
$$\ge (a+\beta)|S| + |T| - (b-\beta)|T|$$
$$= (a+\beta)|S| - (b-\beta-1)|T|$$
$$\ge (a+\beta)\left(\frac{(r-1)(b-\beta)(b+1)}{a+\beta} - 1\right)$$
$$\quad - (b-\beta-1)((r-1)(b+1) - 1)$$
$$= (r-1)(b+1) - 1 - \beta \ge b - \beta$$
$$\ge a \ge 2 \ge \varepsilon(S,T),$$

which is a contradiction. Hence, we have that $|T| \ge (r-1)(b+1)$ if

3.2 Fractional (g, f)-Deleted Graphs

$d_{G-S}(x) = 1$ for any $x \in T$.

In the following, we prove that $|T| \geq (r-1)(b+1) + 1$ in other cases.

Suppose that $|T| \leq (r-1)(b+1)$. We consider three cases by the value of h_1.

Case 1 $h_1 = 0$.

Note that $d_{G-S}(T) \geq \varepsilon(S, T)$. Combining this with (3.2.13) and $r \geq 2$, we obtain

$$\begin{aligned}\delta_G(S, T) &\geq f(S) + d_{G-S}(T) - g(T) \\ &\geq (a + \beta)|S| + \varepsilon(S, T) - (b - \beta)|T| \\ &\geq (r-1)(b - \beta)(b+1) + \varepsilon(S, T) \\ &\quad - (b - \beta)(r-1)(b+1) \\ &= \varepsilon(S, T),\end{aligned}$$

which contradicts (3.2.12).

Case 2 $h_1 = 1$.

In this case, there exists $t \in T$ such that $d_{G-S}(t) \geq 2$. In terms of (3.2.12), (3.2.13) and $r \geq 2$, we have

$$\begin{aligned}\varepsilon(S,T) - 1 &\geq \delta_G(S,T) = f(S) + d_{G-S}(T) - g(T) \\ &\geq (a+\beta)|S| + |T| + 1 - (b-\beta)|T| \\ &= (a+\beta)|S| - (b-\beta-1)|T| + 1 \\ &\geq (a+\beta)\left(\frac{(r-1)(b-\beta)(b+1)}{a+\beta} - 1\right) \\ &\quad - (b-\beta-1)(r-1)(b+1) + 1 \\ &= (r-1)(b+1) - (a+\beta) + 1 \\ &\geq (b+1) - (a+\beta) + 1 \\ &\geq 2 \geq \varepsilon(S,T),\end{aligned}$$

which is a contradiction.

Case 3 $2 \leq h_1 \leq b - \beta$.

Using (3.2.13) and $r \geq 2$, we obtain

$$\delta_G(S,T) \geq f(S) + d_{G-S}(T) - g(T)$$
$$\geq (a+\beta)|S| + h_1|T| - (b-\beta)|T|$$
$$= (a+\beta)|S| - (b-\beta-h_1)|T|$$
$$\geq (a+\beta)\left(\frac{(r-1)(b-\beta)(b+1)}{a+\beta} - h_1\right)$$
$$- (b-\beta-h_1)(r-1)(b+1)$$
$$= h_1((r-1)(b+1) - (a+\beta))$$
$$\geq h_1((b+1) - (a+\beta)) \geq h_1$$
$$\geq 2 \geq \varepsilon(S,T),$$

which contradicts (3.2.12). Hence, we obtain that $|T| \geq (r-1)(b+1) + 1$ in other cases. This completes the proof of Claim 2. □

Claim 3 There exists an independent subset $\{x_1, x_2, \cdots, x_r\}$ in G such that $\{x_1, x_2, \cdots, x_r\} \subseteq T$.

Proof If $d_{G-S}(x) = 1$ for any $x \in T$, then by Claim 2 we have

$$|T| \geq (r-1)(b+1).$$

Note that $b \geq 2$ and $d_{G-S}(x) \leq g(x) \leq b - \beta \leq b$ for each $x \in T$. Thus, we obtain $d_{G-S}(x) \leq b - 1$ for any $x \in T$. Combining this with

$$|T| \geq (r-1)(b+1) \geq (r-1)b + 1,$$

we can take above independent subset $\{x_1, x_2, \cdots, x_r\} \subseteq T$ for $\lambda = r$.

Otherwise, we have $|T| \geq (r-1)(b+1) + 1$ by Claim 2. Combining this with $d_{G-S}(x) \leq g(x) \leq b - \beta \leq b$ for any $x \in T$, we can take above independent subset $\{x_1, x_2, \cdots, x_r\} \subseteq T$ for $\lambda = r$. The proof of Claim 3 is complete. □

Using Claim 3 and the condition of Theorem 3.2.4, we easily obtain

$$\frac{(b-\beta)n}{a+b} \leq \max\{d_G(x_1), d_G(x_2), \cdots, d_G(x_r)\} \leq |S| + h_r,$$

that is,

$$|S| \geq \frac{(b-\beta)n}{a+b} - h_r. \quad (3.2.14)$$

3.2 Fractional (g, f)-Deleted Graphs

Claim 4 $|S| < \dfrac{(b-\beta)n}{a+b}$.

Proof Note that $d_{G-S}(T) \geq \varepsilon(S,T)$. Combining this with (3.2.12) and $|S| + |T| \leq n$, we have

$$\varepsilon(S,T) - 1 \geq \delta_G(S,T) = f(S) + d_{G-S}(T) - g(T)$$
$$\geq (a+\beta)|S| + \varepsilon(S,T) - (b-\beta)|T|$$
$$\geq (a+\beta)|S| + \varepsilon(S,T) - (b-\beta)(n - |S|)$$
$$= (a+b)|S| + \varepsilon(S,T) - (b-\beta)n,$$

which implies

$$|S| < \frac{(b-\beta)n}{a+b}.$$

This completes the proof of Claim 4. $\square$

From (3.2.14) and Claim 4, we obtain $h_r > 0$. In terms of the integrity of h_r, we have

$$h_r \geq 1. \qquad (3.2.15)$$

Note that

$$|N_T[x_i]| - \left|N_T[x_i] \bigcap \left(\bigcup_{j=1}^{i-1} N_T[x_j]\right)\right| \geq 1, \quad 2 \leq i \leq r-1 \qquad (3.2.16)$$

and

$$\left|\bigcup_{j=1}^{i} N_T[x_j]\right| \leq \sum_{j=1}^{i} |N_T[x_j]| \leq \sum_{j=1}^{i} (d_{G-S}(x_j) + 1)$$
$$= \sum_{j=1}^{i}(h_j + 1), \quad 1 \leq i \leq r. \qquad (3.2.17)$$

According to (3.2.12), (3.2.16), (3.2.17), $|S| + |T| \leq n$ and $h_r \leq b - \beta$, we have

$$1 \geq \varepsilon(S,T) - 1 \geq \delta_G(S,T) = f(S) + d_{G-S}(T) - g(T)$$
$$\geq (a+\beta)|S| + d_{G-S}(T) - (b-\beta)|T|$$
$$\geq (a+\beta)|S| + h_1|N_T[x_1]| + h_2\left(|N_T[x_2]| - \left|N_T[x_2]\bigcap N_T[x_1]\right|\right)$$

$$+ \cdots + h_{r-1}\left(|N_T[x_{r-1}]| - \left|N_T[x_{r-1}] \cap \left(\bigcup_{i=1}^{r-2} N_T[x_i]\right)\right|\right)$$

$$+ h_r\left(|T| - \left|\bigcup_{i=1}^{r-1} N_T[x_i]\right|\right) - (b-\beta)|T|$$

$$\geq (a+\beta)|S| + (h_1 - h_r)|N_T[x_1]| + \sum_{i=2}^{r-1} h_i + (h_r - (b-\beta))|T|$$

$$- h_r \sum_{i=2}^{r-1} |N_T[x_i]|$$

$$\geq (a+\beta)|S| + (h_1 - h_r)(h_1 + 1) + \sum_{i=2}^{r-1} h_i + (h_r - (b-\beta))|T|$$

$$- h_r \sum_{i=2}^{r-1}(h_i + 1)$$

$$= (a+\beta)|S| + h_1^2 + \sum_{i=1}^{r-1} h_i - h_r \sum_{i=1}^{r-1}(h_i + 1) + (h_r - (b-\beta))|T|$$

$$\geq (a+\beta)|S| + h_1^2 + \sum_{i=1}^{r-1} h_i - h_r \sum_{i=1}^{r-1}(h_i + 1)$$

$$+ (h_r - (b-\beta))(n - |S|)$$

$$= (a+b-h_r)|S| + h_1^2 + \sum_{i=1}^{r-1} h_i - h_r \sum_{i=1}^{r-1}(h_i + 1)$$

$$+ (h_r - (b-\beta))n,$$

that is,

$$1 \geq (a+b-h_r)|S| + h_1^2 + \sum_{i=1}^{r-1} h_i - h_r \sum_{i=1}^{r-1}(h_i+1) + (h_r - (b-\beta))n. \quad (3.2.18)$$

It follows from (3.2.14), (3.2.15), (3.2.18), $0 \leq h_1 \leq h_2 \leq \cdots \leq h_r \leq b - \beta$ and

$$n \geq \frac{(a+b)(a+b+1+(r-2)(b-\beta))}{a+\beta}$$

that

3.2 Fractional (g, f)-Deleted Graphs

$$1 \geq (a+b-h_r)|S| + h_1^2 + \sum_{i=1}^{r-1} h_i - h_r \sum_{i=1}^{r-1}(h_i + 1)$$
$$+ (h_r - (b-\beta))n$$
$$\geq (a+b-h_r)\left(\frac{(b-\beta)n}{a+b} - h_r\right) + h_1^2 - (h_r - 1)\sum_{i=1}^{r-1} h_i$$
$$- h_r(r-1) + (h_r - (b-\beta))n$$
$$\geq (a+b-h_r)\left(\frac{(b-\beta)n}{a+b} - h_r\right) + h_1^2 - (h_r - 1)(r-1)h_r$$
$$- h_r(r-1) + (h_r - (b-\beta))n$$
$$\geq (a+b-h_r)\left(\frac{(b-\beta)n}{a+b} - h_r\right) - (h_r - 1)(r-1)h_r$$
$$- h_r(r-1) + (h_r - (b-\beta))n$$
$$= h_r\left(\frac{(a+\beta)n}{a+b} - (a+b) - (r-2)h_r\right)$$
$$\geq h_r(a+b+1+(r-2)(b-\beta) - (a+b) - (r-2)(b-\beta))$$
$$= h_r \geq 1.$$

Thus equality holds throughout, which implies that $h_1 = h_2 = \cdots = h_r$ and $h_1 = 0$. Therefore, $h_r = 0$. Which contradicts (3.2.15). This completes the proof of Theorem 3.2.4. □

If $g(x) \equiv f(x)$ in Theorem 3.2.4, then we get immediately the following corollary.

Corollary 6 Let $r \geq 2$ and $2 \leq a \leq b$ be integers, and let G be a graph of order n with
$$n \geq \frac{(a+b)(a+b+1+b(r-2))}{a}$$
and
$$\delta(G) \geq \frac{b(b+1)(r-1)}{a},$$
and let f be an integer-valued function defined on $V(G)$ satisfying $a \leq f(x) \leq b$ for each $x \in V(G)$. If G satisfies

$$\max\{d_G(x_1), d_G(x_2), \cdots, d_G(x_r)\} \geq \frac{bn}{a+b}$$

for any independent subset $\{x_1, x_2, \cdots, x_r\}$ *in G, then G is a fractional f-deleted graph.*

If $a = b = k$ in Theorem 3.2.4, then $n \geq 2rk + 2$. But in this case, the condition on order n can be improved.

Theorem 3.2.5 [39] *Let k and r be two integers with* $k \geq 2$ *and* $r \geq 2$, *and let G be a graph of order n with* $n \geq 2rk + 1$, *and*

$$\delta(G) \geq (r-1)(k+1).$$

If G satisfies

$$\max\{d_G(x_1), d_G(x_2), \cdots, d_G(x_r)\} \geq \frac{n}{2}$$

for any independent subset $\{x_1, x_2, \cdots, x_r\}$ *in G, then G is a fractional k-deleted graph.*

Proof The proof of Theorem 3.2.5 is quite similar to that of Theorem 3.2.4. Suppose that G satisfies the assumption of Theorem 3.2.5, but it is not a fractional k-deleted graph. Then by Theorem 3.1.1, there exists some subset $S \subseteq V(G)$ such that

$$\delta_G(S, T) = k|S| + d_{G-S}(T) - k|T| \leq \varepsilon(S, T) - 1,$$

where $T = \{x : x \in V(G) \setminus S, d_{G-S}(x) \leq k\}$. Under the assumption of Theorem 3.2.5, (3.2.14), (3.2.15) and (3.2.18) also holds when $a = b = k$ and $\beta = 0$.

We now consider the following two cases.

Case 1 n is even.

Since $n \geq 2rk + 1$ and n is even, we have

$$n \geq 2rk + 2. \tag{3.2.19}$$

In terms of (3.2.14), (3.2.15), (3.2.18) and (3.2.19) and $0 \leq h_1 \leq h_2 \leq \cdots \leq h_r \leq k$, we have

3.2 Fractional (g, f)-Deleted Graphs

$$1 \geq (2k - h_r)|S| + h_1^2 + \sum_{i=1}^{r-1} h_i - h_r \sum_{i=1}^{r-1}(h_i + 1) + (h_r - k)n$$

$$= (2k - h_r)|S| + h_1^2 - (h_r - 1)\sum_{i=1}^{r-1} h_i$$

$$\quad - h_r(r-1) + (h_r - k)n$$

$$\geq (2k - h_r)\left(\frac{n}{2} - h_r\right) + h_1^2 - (h_r - 1)\sum_{i=1}^{r-1} h_i$$

$$\quad - h_r(r-1) + (h_r - k)n$$

$$= \frac{n}{2}h_r - (2k + r - 1)h_r - (h_r - 1)\sum_{i=1}^{r-1} h_i + h_r^2 + h_1^2$$

$$\geq \frac{n}{2}h_r - (2k + r - 1)h_r - (h_r - 1)(r - 1)h_r + h_r^2$$

$$= h_r\left(\frac{n}{2} - 2k - (r-2)h_r\right)$$

$$\geq h_r(rk + 1 - 2k - (r-2)k) = h_r \geq 1.$$

Thus equality holds throughout, which implies that $h_1 = h_2 = \cdots = h_r = 1$ and $h_1 = 0$, which is impossible.

Case 2 n is odd.

According to (3.2.14) and the integrity of $|S|$, we have

$$|S| \geq \frac{n+1}{2} - h_r. \tag{3.2.20}$$

From (3.2.15), (3.2.18), (3.2.20), $0 \leq h_1 \leq h_2 \leq \cdots \leq h_r \leq k$ and $n \geq 2rk + 1$, we have

$$1 \geq (2k - h_r)|S| + h_1^2 + \sum_{i=1}^{r-1} h_i - h_r \sum_{i=1}^{r-1}(h_i + 1) + (h_r - k)n$$

$$= (2k - h_r)|S| + h_1^2 - (h_r - 1)\sum_{i=1}^{r-1} h_i - (r-1)h_r + (h_r - k)n$$

$$\geq (2k - h_r)\left(\frac{n+1}{2} - h_r\right) + h_1^2 - (h_r - 1)(r - 1)h_r$$

$$\quad - (r-1)h_r + (h_r - k)n$$

$$= h_r\left(\frac{n-1}{2} - 2k - (r-2)h_r\right) + h_1^2 + k$$

$$\geq h_r(rk - 2k - (r-2)k) + h_1^2 + k$$

$$= h_1^2 + k \geq k \geq 1.$$

Thus equality holds throughout, which implies that $h_1 = h_2 = \cdots = h_r = k = 1$ and $h_1 = 0$, which is impossible. The proof of Theorem 3.2.5 is complete. □

Remark 3.2.3 In Theorem 3.2.4, the lower bound on the condition

$$\max\{d_G(x_1), d_G(x_2), \cdots, d_G(x_r)\} \geq \frac{(b-\beta)n}{a+b}$$

is best possible in the sense that we cannot replace $\frac{(b-\beta)n}{a+b}$ by $\frac{(b-\beta)n}{a+b} - 1$, which is shown in the following example:

We construct a graph $G = K_{rat} \vee (rbt+1)K_1$, where $r \geq 2$, $\beta \geq 0$ and $2 \leq a = b - \beta$ be integers and t is a enough large positive integer. Let $g(x)$ and $f(x)$ be integer-valued functions defined on $V(G)$ satisfying $g(x) = a$ and $f(x) = b = a + \beta$ for any $x \in V(G)$. Clearly,

$$\delta(G) = rat \geq \frac{(r-1)(b-\beta)(b+1)}{a+\beta}$$

and

$$\frac{(b-\beta)n}{a+b} > \max\{d_G(x_1), d_G(x_2), \cdots, d_G(x_r)\}$$

$$= rat > \frac{(b-\beta)n}{a+b} - 1$$

for any independent subset $\{x_1, x_2, \cdots, x_r\}$ in G. Set $S = V(K_{rat})$ and $T = V((rbt+1)K_1)$. Then $|S| = rkt$, $|T| = rkt + 1$, $d_{G-S}(T) = 0$ and $\varepsilon(S,T) = 0$. Thus, we obtain

$$\delta_G(S,T) = f(S) + d_{G-S}(T) - g(T) = b|S| + d_{G-S}(T) - a|T|$$

$$= b(rat) - a(rbt+1) = -a$$

$$< 0 = \varepsilon(S,T).$$

In terms of Theorem 3.1.1, G is not a fractional (g, f)-deleted graph. And so, the lower bound on the condition

3.2 Fractional (g, f)-Deleted Graphs

$$\max\{d_G(x_1), d_G(x_2), \cdots, d_G(x_r)\} \geq \frac{n}{2}$$

in Theorem 3.2.5 is also best possible.

Remark 3.2.4 We show that the bound on n in Theorem 3.2.5 is also best possible. Let $r = 2$ and $k \geq 2$. We construct a graph

$$G = K_{rk-1} \vee \left(K_1 \cup \frac{rk}{2} K_2\right).$$

Then G satisfies all conditions of Theorem 3.2.5 except $n = 2rk$. Let $S = V(K_{rk-1})$ and $T = V\left(K_1 \cup \frac{rk}{2} K_2\right)$. Clearly, $d_{G-S}(T) = rk$ and T is not independent. In this case, $\varepsilon(S, T) = 2$. Thus, we have

$$\delta_G(S, T) = k|S| + d_{G-S}(T) - k|T|$$
$$= k(rk - 1) + rk - k(rk + 1)$$
$$= 0 < 2 = \varepsilon(S, T).$$

According to Theorem 3.1.1, G is not a fractional k-deleted graph.

Chapter 4
Fractional (k, m)-Deleted Graphs

In this chapter, we study fractional (k, m)-deleted graphs which are the generalizations of fractional k-factors and fractional k-deleted graphs. We first give a criterion for a graph to be a fractional (k, m)-deleted graph. Then we use the criterion to obtain some graphic parameter (such as minimum degree, neighborhood, toughness and binding number, etc.) conditions for graphs to be fractional (k, m)-deleted graphs.

4.1 A Criterion for Fractional (k, m)-Deleted Graphs

Let G be a graph, $k \geq 1$ be an integer and $h : E(G) \to [0, 1]$ be a function. If $\sum_{e \ni x} h(e) = k$ holds for any $x \in V(G)$, then we call $G[F_h]$ a **fractional k-factor** of G with indicator function h where

$$F_h = \{e \in E(G) : h(e) > 0\}.$$

A graph G is called a **fractional (k, m)-deleted graph** if there exists a fractional k-factor $G[F_h]$ of G with indicator function h such that $h(e) = 0$ for any $e \in E(H)$, where H is any subgraph of G with m edges. A fractional (k, m)-deleted graph is simply called a **fractional k-deleted graph** if $m = 1$. If $m = 0$, then a graph G admits a fractional k-factor.

For the convenience of reference, we restate Fractional k-Factor Theorem (i.e., Theorem 2.1.1) below.

Theorem 4.1.1 [12]   *Let G be a graph. Then G has a fractional k-factor*

4.1 A Criterion for Fractional (k, m)-Deleted Graphs

if and only if for every subset S of $V(G)$,

$$\delta_G(S, T) = k|S| + d_{G-S}(T) - k|T| \geq 0,$$

where $T = \{x : x \in V(G) \setminus S, d_{G-S}(x) \leq k - 1\}$.

A necessary and sufficient condition for a graph to be a fractional (k, m)-deleted graph was obtained by Zhou[26] in 2009, which is the following theorem.

Theorem 4.1.2 [26] Let $k \geq 1$ and $m \geq 0$ be two integers, and let G be a graph and H a subgraph of G with m edges. Then G is a fractional (k, m)-deleted graph if and only if for any subset S of $V(G)$,

$$\delta_G(S, T) = k|S| + \sum_{x \in T} d_{G-S}(x) - k|T| \geq \sum_{x \in T} d_H(x) - e_H(S, T),$$

where $T = \{x : x \in V(G) \setminus S, d_{G-S}(x) - d_H(x) + e_H(x, S) \leq k - 1\}$.

Proof Let $G' = G - E(H)$. Then G is a fractional (k, m)-deleted graph if and only if G' has a fractional k-factor. According to Theorem 4.1.1, this is true if and only if for any subset S of $V(G)$,

$$\delta_{G'}(S, T') = k|S| + d_{G'-S}(T') - k|T'| \geq 0,$$

where $T' = \{x : x \in V(G) \setminus S, d_{G'-S}(x) \leq k - 1\}$.

It is easy to see that

$$d_{G'-S}(x) = d_{G-S}(x) - d_H(x) + e_H(x, S)$$

for any $x \in T'$. By the definitions of T' and T, we have $T' = T$. Hence, we obtain

$$\delta_{G'}(S, T') = \delta_G(S, T) - \sum_{x \in T} d_H(x) + e_H(S, T).$$

Thus, $\delta_{G'}(S, T') \geq 0$ if and only if

$$\delta_G(S, T) \geq \sum_{x \in T} d_H(x) - e_H(S, T).$$

It follows that G is a fractional (k, m)-deleted graph if and only if

$$\delta_G(S,T) = k|S| + \sum_{x \in T} d_{G-S}(x) - k|T| \geq \sum_{x \in T} d_H(x) - e_H(S,T).$$

The proof of Theorem 4.1.2 is complete. □

4.2 Degree Conditions for Fractional (k,m)-Deleted Graphs

In this section, we discuss the existence of fractional (k,m)-deleted graphs in terms of degree conditions. We first show a minimum degree condition for the existence of fractional (k,m)-deleted graphs, which is the following theorem.

Theorem 4.2.1 [26] *Let $k \geq 1$ and $m \geq 1$ be two integers. Let G be a graph of order n with $n \geq 4k - 5 + 2(2k+1)m$. If*

$$\delta(G) \geq \frac{n}{2},$$

then G is a fractional (k,m)-deleted graph.

Proof Suppose that G satisfies the assumption of the theorem, but is not a fractional (k,m)-deleted graph. Then by Theorem 4.1.2, there exists some subset S of $V(G)$ such that

$$k|S| + \sum_{x \in T}(d_{G-S}(x) - d_H(x) + e_H(x,S) - k) \leq -1, \qquad (4.2.1)$$

where $T = \{x : x \in V(G) \setminus S, d_{G-S}(x) - d_H(x) + e_H(x,S) \leq k - 1\}$.

At first, we prove the following claims.

Claim 1 $|S| \geq 1$.

Proof If $S = \emptyset$, then by (4.2.1), we have

$$-1 \geq \sum_{x \in T}(d_G(x) - d_H(x) - k) \geq \sum_{x \in T}(\delta(G) - m - k) \geq 0,$$

which is a contradiction. □

Claim 2 $|T| \geq k + 1$.

4.2 Degree Conditions for Fractional (k, m)-Deleted Graphs

Proof If $|T| \leq k$, then by (4.2.1), Claim 1 and $\delta(G) \geq \frac{n}{2}$, we have

$$-1 \geq k|S| + \sum_{x \in T}(d_{G-S}(x) - d_H(x) + e_H(x, S) - k)$$

$$\geq |T||S| + \sum_{x \in T}(d_{G-S}(x) - d_H(x) + e_H(x, S) - k)$$

$$= \sum_{x \in T}(|S| + d_{G-S}(x) - d_H(x) + e_H(x, S) - k)$$

$$\geq \sum_{x \in T}(d_G(x) - d_H(x) + e_H(x, S) - k)$$

$$\geq \sum_{x \in T}(\delta(G) - m - k)$$

$$\geq 0,$$

which is a contradiction. □

According to Claim 2, we have $T \neq \emptyset$. Thus, we may define

$$h = \min\{d_{G-S}(x) - d_H(x) + e_H(x, S) : x \in T\}.$$

And let x_1 be a vertex in T satisfying

$$d_{G-S}(x_1) - d_H(x_1) + e_H(x_1, S) = h.$$

Then we have $0 \leq h \leq k - 1$ according to the definition of T and

$$d_G(x_1) \leq d_{G-S}(x_1) + |S| = h + d_H(x_1) - e_H(x_1, S) + |S|.$$

In view of the condition of Theorem 4.2.1, the following inequalities hold:

$$\frac{n}{2} \leq \delta(G) \leq d_G(x_1) \leq h + d_H(x_1) - e_H(x_1, S) + |S|,$$

that is,

$$|S| \geq \frac{n}{2} - (h + d_H(x_1) - e_H(x_1, S)). \quad (4.2.2)$$

Now in order to prove the theorem, we shall deduce some contradictions in view of the following two cases.

Case 1 $h = 0$.

By (4.2.1), (4.2.2) and $|S| + |T| \leq n$, we get

$$-1 \geq k|S| + \sum_{x \in T}(d_{G-S}(x) - d_H(x) + e_H(x, S) - k)$$

$$\geq k|S| + h|T| - k|T| = k|S| - k|T|$$

$$\geq k|S| - k(n - |S|) = 2k|S| - kn$$

$$\geq 2k\left(\frac{n}{2} - (d_H(x_1) - e_H(x_1, S))\right) - kn$$

$$= -2k(d_H(x_1) - e_H(x_1, S)),$$

which implies

$$d_H(x_1) - e_H(x_1, S) \geq \frac{1}{2k} > 0.$$

According to the integrity of $d_H(x_1) - e_H(x_1, S)$, we have

$$d_H(x_1) - e_H(x_1, S) \geq 1.$$

For some $x \in T \setminus \{x_1\}$, if

$$d_{G-S}(x) - d_H(x) + e_H(x, S) = 0,$$

then we similarly get $d_H(x) - e_H(x, S) \geq 1$. Hence, one of (1) and (2) holds for any $x \in T \setminus \{x_1\}$:

(1) $d_{G-S}(x) - d_H(x) + e_H(x, S) \geq 1$, or

(2) $d_{G-S}(x) - d_H(x) + e_H(x, S) = 0$ and $d_H(x) - e_H(x, S) \geq 1$.

Thus, we have

$$\sum_{x \in T}(d_{G-S}(x) - d_H(x) + e_H(x, S)) \geq |T| - 2m. \qquad (4.2.3)$$

In view of (4.2.1), (4.2.2), (4.2.3), $h = 0$, $|S| + |T| \leq n$ and $n \geq 4k - 5 + 2(2k+1)m$, we get

$$-1 \geq k|S| + \sum_{x \in T}(d_{G-S}(x) - d_H(x) + e_H(x, S) - k)$$

$$\geq k|S| + |T| - 2m - k|T| = k|S| - (k-1)|T| - 2m$$

$$\geq k|S| - (k-1)(n - |S|) - 2m$$

$$= (2k-1)|S| - (k-1)n - 2m$$

$$\geq (2k-1)\left(\frac{n}{2} - (d_H(x_1) - e_H(x_1, S))\right) - (k-1)n - 2m$$

4.2 Degree Conditions for Fractional (k, m)-Deleted Graphs

$$\geq (2k-1)\left(\frac{n}{2} - m\right) - (k-1)n - 2m = \frac{n}{2} - (2k+1)m$$

$$\geq \frac{4k - 5 + 2(2k+1)m}{2} - (2k+1)m$$

$$> 2k - 3 \geq -1,$$

a contradiction.

Case 2 $1 \leq h \leq k - 1$.

According to (4.2.1), (4.2.2), $n \geq 4k-5+2(2k+1)m$ and $|S|+|T| \leq n$, we obtain

$$-1 \geq k|S| + \sum_{x \in T}(d_{G-S}(x) - d_H(x) + e_H(x, S) - k)$$

$$\geq k|S| + h|T| - k|T| = k|S| - (k-h)|T|$$

$$\geq k|S| - (k-h)(n - |S|) = (2k-h)|S| - (k-h)n$$

$$\geq (2k-h)\left(\frac{n}{2} - (h + d_H(x_1) - e_H(x_1, S))\right) - (k-h)n$$

$$= \frac{hn}{2} - (2k-h)(h + d_H(x_1) - e_H(x_1, S))$$

$$\geq \frac{hn}{2} - (2k-h)(h + m)$$

$$\geq \frac{h(4k - 5 + 2(2k+1)m)}{2} - (2k-h)(h + m)$$

$$> \frac{h(4k - 6 + 2(2k+1)m)}{2} - (2k-h)(h + m)$$

$$= h^2 + 2(k+1)mh - 3h - 2km,$$

that is,

$$-1 > h^2 + 2(k+1)mh - 3h - 2km. \tag{4.2.4}$$

Let $f(h) = h^2 + 2(k+1)mh - 3h - 2km$. Clearly, the function $f(h)$ attains its minimum value at $h = 1$ since $1 \leq h \leq k - 1$. Then we get

$$f(h) \geq f(1).$$

Combining this with (4.2.4) and $m \geq 1$, we have

$$-1 > f(h) \geq f(1) = 2m - 2 \geq 0,$$

which is a contradiction. This completes the proof of Theorem 4.2.1. □

Remark 4.2.1 Let us show that the condition $\delta(G) \geq \frac{n}{2}$ in Theorem 4.2.1 can not be replaced by $\delta(G) \geq \frac{n-1}{2}$. Let

$$G = K_{2k-3+(2k+1)m} \bigvee (((2k-1)m + 2k - 2)K_1 \cup (mK_2)).$$

Then we have $n = 4k - 5 + 2(2k+1)m$ and

$$\delta(G) = 2k - 3 + (2k+1)m = \frac{n-1}{2}.$$

Let $G' = G - E(mK_2)$, $S = V(K_{2k-3+(2k+1)m}) \subseteq V(G)$ and

$$T = V(((2k-1)m + 2k - 2)K_1 \cup (mK_2)) \subseteq V(G),$$

then $|S| = 2k-3+(2k+1)m$, $|T| = 2k-2+(2k+1)m$ and $d_{G'-S}(T) = 0$. Thus, we get

$$\begin{aligned}\delta_{G'}(S,T) &= k|S| + d_{G'-S}(T) - k|T| \\ &= k(2k - 3 + (2k+1)m) - k(2k - 2 + (2k+1)m) \\ &= -k < 0.\end{aligned}$$

By Theorem 4.1.1, G' has no fractional k-factor. Hence, G is not a fractional (k,m)-deleted graph. In the above sense, the result in Theorem 4.2.1 is best possible.

In Theorem 4.2.1, if $m = 1$, then we get the following result.

Theorem 4.2.2 Let $k \geq 1$ be an integer. Let G be a graph of order n with $n \geq 8k - 3$. If

$$\delta(G) \geq \frac{n}{2},$$

then G is a fractional k-deleted graph.

Yu, Liu, Ma and Cao obtained a degree condition for the existence of fractional k-factors.

Theorem 4.2.3 [23] Let k be an integer with $k \geq 1$, and let G be a connected graph of order n with $n \geq 4k - 3$, $\delta(G) \geq k$. If

4.2 Degree Conditions for Fractional (k,m)-Deleted Graphs

$$\max\{d_G(x), d_G(y)\} \geq \frac{n}{2}$$

for each pair of nonadjacent vertices x, y of G, then G has a fractional k-factor.

In the following we give a new degree condition for a graph to a fractional (k,m)-deleted graph, which is an extension of Theorem 4.2.3 and an improvement of Theorem 4.2.1.

Theorem 4.2.4 [35] *Let $k \geq 1$ and $m \geq 0$ be two integers. Let G be a connected graph of order n with*

$$n \geq 4k - 3 + 2(2k+1)m, \quad \delta(G) \geq k + m + \frac{(m+1)^2 - 3}{4k}.$$

If

$$\max\{d_G(x), d_G(y)\} \geq \frac{n}{2}$$

for each pair of nonadjacent vertices x, y of G, then G is a fractional (k,m)-deleted graph.

Proof According to Theorem 4.2.3, the theorem is trivial for $m = 0$. In the following, we consider $m \geq 1$. The proof is by contradiction. We assume that a graph G satisfies the assumption of the theorem, but it is not a fractional (k,m)-deleted graph. Then according to Theorem 4.1.2, there exists some subset S of $V(G)$ such that

$$k|S| + \sum_{x \in T}(d_{G-S}(x) - d_H(x) + e_H(x, S) - k) \leq -1, \quad (4.2.5)$$

where $T = \{x : x \in V(G) \setminus S, d_{G-S}(x) - d_H(x) + e_H(x, S) \leq k - 1\}$ and H is any subgraph of G with m edges.

At first, we prove the following claims.

Claim 1 $|S| + d_{G-S}(x) - d_H(x) + e_H(x, S) - k \geq 0$ for each $x \in V(G)$.

Proof According to $d_H(x) \leq m$ and

$$\delta(G) \geq k + m + \frac{(m+1)^2 - 3}{4k},$$

we get
$$|S| + d_{G-S}(x) - d_H(x) + e_H(x,S) - k$$
$$\geq d_G(x) - m - k \geq \delta(G) - m - k$$
$$\geq \frac{(m+1)^2 - 3}{4k} \geq \frac{-2}{4k} > -1$$

for each $x \in V(G)$.

In terms of the integrity of $|S| + d_{G-S}(x) - d_H(x) + e_H(x,S) - k$, we have
$$|S| + d_{G-S}(x) - d_H(x) + e_H(x,S) - k \geq 0$$

for each $x \in V(G)$. □

Claim 2 $|T| \geq k + 1$.

Proof If $|T| \leq k$, then by (4.2.5) and Claim 1, we obtain
$$-1 \geq k|S| + \sum_{x \in T}(d_{G-S}(x) - d_H(x) + e_H(x,S) - k)$$
$$\geq |T||S| + \sum_{x \in T}(d_{G-S}(x) - d_H(x) + e_H(x,S) - k)$$
$$= \sum_{x \in T}(|S| + d_{G-S}(x) - d_H(x) + e_H(x,S) - k)$$
$$\geq 0,$$

a contradiction. □

In view of Claim 2, $T \neq \emptyset$. Then we may choose a vertex $x_1 \in T$ with
$$h_1 = \min\{d_{G-S}(x) - d_H(x) + e_H(x,S) : x \in T\}$$
$$= d_{G-S}(x_1) - d_H(x_1) + e_H(x_1, S)$$

and $d_H(x_1) - e_H(x_1, S)$ minimum. Further, if $T \setminus N_T[x_1] \neq \emptyset$, we may choose a vertex $x_2 \in T \setminus N_T[x_1]$ with
$$h_2 = \min\{d_{G-S}(x) - d_H(x) + e_H(x,S) : x \in T \setminus N_T[x_1]\}$$
$$= d_{G-S}(x_2) - d_H(x_2) + e_H(x_2, S)$$

and $d_H(x_2) - e_H(x_2, S)$ minimum. Then we have $0 \leq h_1 \leq h_2 \leq k - 1$ and

4.2 Degree Conditions for Fractional (k, m)-Deleted Graphs

$$d_G(x_i) \leq |S| + d_{G-S}(x_i) = |S| + h_i + d_H(x_i) - e_H(x_i, S)$$

for $i = 1, 2$.

According to the choice of x_1, x_2, we have $x_1 x_2 \notin E(G)$. Thus, from the condition of Theorem 4.2.4, the following inequalities hold:

$$\frac{n}{2} \leq \max\{d_G(x_1), d_G(x_2)\}$$

$$\leq |S| + h_2 + \max\{d_H(x_1) - e_H(x_1, S), d_H(x_2) - e_H(x_2, S)\},$$

that is,

$$|S| \geq \frac{n}{2} - h_2 - \max\{d_H(x_1) - e_H(x_1, S), d_H(x_2) - e_H(x_2, S)\}.$$

(4.2.6)

Now in order to prove the theorem, we shall deduce some contradictions in view of the following two cases.

Case 1 $T = N_T[x_1]$.

Clearly, the following inequalities hold by $d_H(x_1) \leq m$:

$$|T| = |N_T[x_1]| \leq d_{G-S}(x_1) + 1$$
$$= h_1 + d_H(x_1) - e_H(x_1, S) + 1$$
$$\leq h_1 + m + 1. \qquad (4.2.7)$$

Since

$$d_G(x_1) \leq |S| + h_1 + d_H(x_1) - e_H(x_1, S)$$

and $d_H(x_1) \leq m$, then we get

$$|S| \geq d_G(x_1) - h_1 - d_H(x_1) + e_H(x_1, S)$$
$$\geq \delta(G) - h_1 - m. \qquad (4.2.8)$$

According to (4.2.5), (4.2.7), (4.2.8) and $0 \leq h_1 \leq k - 1$, we obtain

$$-1 \geq k|S| + \sum_{x \in T}(d_{G-S}(x) - d_H(x) + e_H(x, S) - k)$$

$$\geq k|S| + (h_1 - k)|T|$$

$$\geq k(\delta(G) - h_1 - m) + (h_1 - k)(h_1 + m + 1)$$

$$\geq k\left(k+m+\frac{(m+1)^2-3}{4k}-h_1-m\right)$$
$$+(h_1-k)(h_1+m+1)$$
$$=h_1^2-(2k-m-1)h_1+k^2-(m+1)k$$
$$+\frac{(m+1)^2-3}{4}$$
$$=\left(h_1-k+\frac{m+1}{2}\right)^2-\frac{3}{4}$$
$$\geq -\frac{3}{4}>-1,$$

which is a contradiction.

Case 2 $T\setminus N_T[x_1]\neq\emptyset$.

Subcase 2.1 $h_2=0$.

obviously, $h_1=0$. According to the choice of x_1 and x_2, we have
$$d_H(x_2)-e_H(x_2,S)\geq d_H(x_1)-e_H(x_1,S)\geq 0.$$

Complying this with (4.2.6), we obtain
$$|S|\geq \frac{n}{2}-d_H(x_2)+e_H(x_2,S). \tag{4.2.9}$$

In view of (4.2.5), (4.2.9) and $|S|+|T|\leq n$, we have
$$-1\geq k|S|+\sum_{x\in T}(d_{G-S}(x)-d_H(x)+e_H(x,S)-k)$$
$$\geq k|S|-k|T|\geq k|S|-k(n-|S|)=2k|S|-kn$$
$$\geq 2k\left(\frac{n}{2}-(d_H(x_2)-e_H(x_2,S))\right)-kn$$
$$=-2k(d_H(x_2)-e_H(x_2,S)),$$

which implies
$$d_H(x_2)-e_H(x_2,S)\geq \frac{1}{2k}>0.$$

By the integrity of $d_H(x_2)-e_H(x_2,S)$, we get
$$d_H(x_2)-e_H(x_2,S)\geq 1.$$

4.2 Degree Conditions for Fractional (k, m)-Deleted Graphs

According to the choice of x_1 and x_2, one of (1) and (2) holds for any $u \in T \setminus (\{x_1, x_2\} \cup N_H(\{x_1, x_2\}))$:

(1) $d_{G-S}(u) - d_H(u) + e_H(u, S) \geq 1$, or

(2) $d_{G-S}(u) - d_H(u) + e_H(u, S) = 0$ and $d_H(u) - e_H(u, S) \geq 1$.

Since $\{x_1, x_2\} \cap V(H) \neq \emptyset$ and any vertex $v \in T \setminus (\{x_1, x_2\} \cup V(H))$ satisfies (1), we obtain

$$\sum_{x \in T}(d_{G-S}(x) - d_H(x) + e_H(x, S))$$

$$\geq |T| - 2 - 2m + 1 = |T| - 2m - 1. \qquad (4.2.10)$$

In view of (4.2.5), (4.2.9), (4.2.10), $|S| + |T| \leq n$, $d_H(x_2) \leq m$ and $n \geq 4k - 3 + 2(2k + 1)m$, we have

$$-1 \geq k|S| + \sum_{x \in T}(d_{G-S}(x) - d_H(x) + e_H(x, S) - k)$$

$$\geq k|S| + |T| - 2m - 1 - k|T| = k|S| - (k-1)|T| - 2m - 1$$

$$\geq k|S| - (k-1)(n - |S|) - 2m - 1$$

$$= (2k-1)|S| - (k-1)n - 2m - 1$$

$$\geq (2k-1)\left(\frac{n}{2} - d_H(x_2) + e_H(x_2, S)\right) - (k-1)n - 2m - 1$$

$$\geq (2k-1)\left(\frac{n}{2} - m\right) - (k-1)n - 2m - 1$$

$$= \frac{n}{2} - (2k+1)m - 1$$

$$\geq \frac{4k - 3 + 2(2k+1)m}{2} - (2k+1)m - 1$$

$$= \frac{4k - 3}{2} - 1 > 2k - 3 \geq -1,$$

this is a contradiction.

Subcase 2.2 $1 \leq h_2 \leq k - 1$.

By $d_H(x_1) \leq m$, we have

$$|N_T[x_1]| \leq d_{G-S}(x_1) + 1 = h_1 + d_H(x_1) - e_H(x_1, S) + 1$$

$$\leq h_1 + d_H(x_1) + 1 \leq h_1 + m + 1.$$

Since $|E(H)| = m$ and $x_1x_2 \notin E(G)$, then $d_H(x_1) + d_H(x_2) \leq m$. Complying the above inequality with (4.2.6), we obtain

$$|S| \geq \frac{n}{2} - h_2 - (d_H(x_1) + d_H(x_2)) \geq \frac{n}{2} - h_2 - m. \qquad (4.2.11)$$

According to (4.2.5), (4.2.11), $m \geq 1$, $|N_T[x_1]| \leq h_1 + m + 1$, $0 \leq h_1 \leq h_2 \leq k-1$, $n \geq 4k - 3 + 2(2k+1)m$ and $|S| + |T| \leq n$, we get

$$-1 \geq k|S| + \sum_{x \in T}(d_{G-S}(x) - d_H(x) + e_H(x, S) - k)$$

$$\geq k|S| + h_1|N_T[x_1]| + h_2(|T| - |N_T[x_1]|) - k|T|$$

$$= k|S| + (h_1 - h_2)|N_T[x_1]| + (h_2 - k)|T|$$

$$\geq k|S| + (h_1 - h_2)(h_1 + m + 1) + (h_2 - k)(n - |S|)$$

$$= (2k - h_2)|S| + (h_1 - h_2)(h_1 + m + 1) - (k - h_2)n$$

$$\geq (2k - h_2)\left(\frac{n}{2} - h_2 - m\right) + (h_1 - h_2)(h_1 + m + 1) - (k - h_2)n$$

$$= \frac{n}{2}h_2 - (2k - h_2)(h_2 + m) + (h_1 - h_2)(h_1 + m + 1)$$

$$\geq \left(2k - \frac{3}{2} + (2k+1)m\right)h_2 - (2k - h_2)(h_2 + m)$$
$$+ (h_1 - h_2)(h_1 + m + 1)$$

$$= h_2^2 + \left((2k+1)m - \frac{5}{2}\right)h_2 - h_2 h_1 + h_1^2 + (m+1)h_1 - 2km$$

$$= \frac{3}{4}h_2^2 + \left((2k+1)m - \frac{5}{2}\right)h_2 + \left(\frac{1}{2}h_2 - h_1\right)^2$$
$$+ (m+1)h_1 - 2km$$

$$\geq \frac{3}{4}h_2^2 + \left((2k+1)m - \frac{5}{2}\right)h_2 - 2km$$

$$\geq \frac{3}{4} + \left((2k+1)m - \frac{5}{2}\right) - 2km$$

$$= m - \frac{7}{4} \geq -\frac{3}{4} > -1.$$

It is a contradiction. This completes the proof of Theorem 4.2.4. $\square$

Remark 4.2.2 In Theorem 4.2.4, the bound in the assumption

$$\max\{d_G(x), d_G(y)\} \geq \frac{n}{2}$$

is best possible in the sense that we cannot replace $\frac{n}{2}$ by $\frac{n}{2} - 1$. We can show this by constructing a graph

$$G = ((kt - 2m)K_1 \cup mK_2) \vee (kt + 1)K_1,$$

where $k \geq 1$, $m \geq 0$ and $t \geq 2(m+1) + \frac{m-2}{k}$ are three integers. Then it follows that

$$|V(G)| = n = 2kt + 1 \geq 4k - 3 + 2(2k+1)m$$

and

$$\frac{n}{2} > \max\{d_G(x), d_G(y)\} > \frac{n}{2} - 1$$

for each pair of nonadjacent vertices x, y of $(kt+1)K_1 \subset G$. Let

$$G' = G - E(mK_2), \quad S = V((kt - 2m)K_1 \cup mK_2)) \subseteq V(G)$$

and $T = V((kt+1)K_1) \subseteq V(G)$. Then $|S| = kt$, $|T| = kt + 1$ and $d_{G'-S}(T) = 0$. Thus, we get

$$\delta_{G'}(S, T) = k|S| + d_{G'-S}(T) - k|T|$$
$$= k^2 t - k(kt + 1) = -k < 0$$

By Theorem 4.1.1, G' has no fractional k-factor, that is, G is not a fractional (k, m)-deleted graph. In the above sense, the result in Theorem 4.2.4 is best possible.

4.3 Neighborhood and Fractional (k, m)-Deleted Graphs

In this section, we consider the neighborhood conditions to guarantee a graph to be a fractional (k, m)-deleted graph. We obtained two results on fractional (k, m)-deleted graphs by using the neighborhood conditions.[27, 28]

Theorem 4.3.1 [27] *Let $k \geq 2$ and $m \geq 0$ be two integers. Let G be a connected graph of order n with*

Chapter 4 Fractional (k,m)-Deleted Graphs

$$n \geq 9k - 1 - 4\sqrt{2(k-1)^2 + 2} + 2(2k+1)m,$$

$$\delta(G) \geq k + m + \frac{(m+1)^2 - 1}{4k}.$$

If

$$|N_G(x) \cup N_G(y)| \geq \frac{1}{2}(n + k - 2)$$

for each pair of nonadjacent vertices x, y of G, then G is a fractional (k,m)-deleted graph.

In order to prove Theorem 4.3.1, we depend on the following lemma.

Lemma 4.3.1 [7] Let k be an integer such that $k \geq 1$. Then

$$9k - 1 - 4\sqrt{2(k-1)^2 + 2} \begin{cases} > 3k + 5, & \text{for } k \geq 4, \\ > 3k + 4, & \text{for } k = 3, \\ = 3k + 3, & \text{for } k = 2, \\ > 2, & \text{for } k = 1. \end{cases}$$

Proof of Theorem 4.3.1 According to Theorem 2.1.6, the theorem is trivial for $m = 0$. In the following, we consider $m \geq 1$.

Suppose that G satisfies the conditions of Theorem 4.3.1, but is not a fractional (k,m)-deleted graph. From Theorem 4.1.2 there exists a subset S of $V(G)$ such that

$$k|S| + \sum_{x \in T}(d_{G-S}(x) - d_H(x) + e_H(x, S) - k) \leq -1, \quad (4.3.1)$$

where $T = \{x : x \in V(G) \setminus S, d_{G-S}(x) - d_H(x) + e_H(x, S) \leq k - 1\}$ and H is any subgraph of G with m edges.

At first, we prove the following claims.

Claim 1 $|S| \geq 1$.

Proof If $S = \emptyset$, then according to (4.3.1), $d_H(x) \leq m$ and

$$\delta(G) \geq k + m + \frac{(m+1)^2 - 1}{4k},$$

we get

4.3 Neighborhood and Fractional (k,m)-Deleted Graphs

$$-1 \geq \sum_{x \in T}(d_G(x) - d_H(x) - k) \geq \sum_{x \in T}(\delta(G) - m - k)$$

$$\geq \sum_{x \in T}\frac{(m+1)^2 - 1}{4k} \geq 0,$$

this is a contradiction. □

Claim 2 $|T| \geq k+1$.

Proof If $|T| \leq k$, then by (4.3.1), Claim 1, $d_H(x) \leq m$ and

$$\delta(G) \geq k + m + \frac{(m+1)^2 - 1}{4k},$$

we obtain

$$-1 \geq k|S| + \sum_{x \in T}(d_{G-S}(x) - d_H(x) + e_H(x,S) - k)$$

$$\geq |T||S| + \sum_{x \in T}(d_{G-S}(x) - d_H(x) + e_H(x,S) - k)$$

$$= \sum_{x \in T}(|S| + d_{G-S}(x) - d_H(x) + e_H(x,S) - k)$$

$$\geq \sum_{x \in T}(d_G(x) - d_H(x) + e_H(x,S) - k)$$

$$\geq \sum_{x \in T}(\delta(G) - m - k) \geq \sum_{x \in T}\frac{(m+1)^2 - 1}{4k}$$

$$\geq 0,$$

a contradiction. □

From Claim 2, $T \neq \emptyset$. Let

$$h_1 = \min\{d_{G-S}(x) - d_H(x) + e_H(x,S) : x \in T\},$$

and choose $x_1 \in T$ with

$$d_{G-S}(x_1) - d_H(x_1) + e_H(x_1, S) = h_1$$

and $d_H(x_1) - e_H(x_1, S)$ is minimum. Further, if $T \setminus N_T[x_1] \neq \emptyset$, we define

$$h_2 = \min\{d_{G-S}(x) - d_H(x) + e_H(x,S) : x \in T \setminus N_T[x_1]\},$$

and choose $x_2 \in T \setminus N_T[x_1]$ with

$$d_{G-S}(x_2) - d_H(x_2) + e_H(x_2, S) = h_2$$

and $d_H(x_2) - e_H(x_2, S)$ is minimum. Then we obtain $0 \leq h_1 \leq h_2 \leq k-1$ by the definition of T.

In view of the choice of x_1, x_2, we have $x_1 x_2 \notin E(G)$. Thus, by the condition of Theorem 4.3.1, the following inequalities hold:

$$\frac{n+k-2}{2} \leq |N_G(x_1) \cup N_G(x_2)|$$
$$\leq d_{G-S}(x_1) + d_{G-S}(x_2) + |S|$$
$$= |S| + h_1 + d_H(x_1) - e_H(x_1, S) + h_2$$
$$+ d_H(x_2) - e_H(x_2, S),$$

which implies

$$|S| \geq \frac{n+k-2}{2} - (h_1 + h_2 + d_H(x_1) + d_H(x_2)$$
$$- e_H(x_1, S) - e_H(x_2, S)). \qquad (4.3.2)$$

Now in order to prove the theorem, we shall deduce some contradictions according to the following two cases.

Case 1 $T = N_T[x_1]$.

Clearly, the following inequalities hold by $d_H(x_1) \leq m$:

$$|T| = |N_T[x_1]| \leq d_{G-S}(x_1) + 1$$
$$= h_1 + d_H(x_1) - e_H(x_1, S) + 1$$
$$\leq h_1 + m + 1. \qquad (4.3.3)$$

In view of

$$\delta(G) \leq d_G(x_1) \leq |S| + d_{G-S}(x_1)$$
$$= |S| + h_1 + d_H(x_1) - e_H(x_1, S)$$

and $d_H(x_1) \leq m$, then we have

$$|S| \geq \delta(G) - h_1 - d_H(x_1) + e_H(x_1, S)$$
$$\geq \delta(G) - h_1 - m. \qquad (4.3.4)$$

By (4.3.1), (4.3.3), (4.3.4) and $0 \leq h_1 \leq k-1$, we get

4.3 Neighborhood and Fractional (k, m)-Deleted Graphs

$$-1 \geq k|S| + \sum_{x \in T}(d_{G-S}(x) - d_H(x) + e_H(x, S) - k)$$

$$\geq k|S| + (h_1 - k)|T|$$

$$\geq k(\delta(G) - h_1 - m) + (h_1 - k)(h_1 + m + 1)$$

$$\geq k\left(k + m + \frac{(m+1)^2 - 1}{4k} - h_1 - m\right)$$

$$+ (h_1 - k)(h_1 + m + 1)$$

$$= h_1^2 - (2k - m - 1)h_1 + k^2 - (m+1)k + \frac{(m+1)^2 - 1}{4}$$

$$= \left(h_1 - k + \frac{m+1}{2}\right)^2 - \frac{1}{4}$$

$$\geq -\frac{1}{4} > -1.$$

This is a contradiction.

Case 2 $T \setminus N_T[x_1] \neq \emptyset$.

From $|E(H)| = m$ and $x_1 x_2 \notin E(G)$, we get

$$d_H(x_1) + d_H(x_2) \leq m. \qquad (4.3.5)$$

Subcase 2.1 $h_2 = 0$.

Clearly, $h_1 = 0$. By (4.3.1), (4.3.2), and $|S| + |T| \leq n$, we obtain

$$-1 \geq k|S| + \sum_{x \in T}(d_{G-S}(x) - d_H(x) + e_H(x, S) - k)$$

$$\geq k|S| - k|T| \geq k|S| - k(n - |S|) = 2k|S| - kn$$

$$\geq 2k\left(\frac{n+k-2}{2} - (d_H(x_1) + d_H(x_2) - e_H(x_1, S) - e_H(x_2, S))\right)$$

$$- kn$$

$$= k^2 - 2k - 2k(d_H(x_1) + d_H(x_2) - e_H(x_1, S) - e_H(x_2, S)),$$

that is,

$$d_H(x_1) + d_H(x_2) - e_H(x_1, S) - e_H(x_2, S) \geq \frac{k^2 - 2k + 1}{2k} > 0.$$

According to the integrity of $d_H(x_1) + d_H(x_2) - e_H(x_1, S) - e_H(x_2, S)$,

we have
$$d_H(x_1) + d_H(x_2) - e_H(x_1, S) - e_H(x_2, S) \geq 1.$$

In view of the choice of x_1 and x_2, one of (1) and (2) holds for any $u \in T \setminus (\{x_1, x_2\} \cup N_H(\{x_1, x_2\}))$:

(1) $d_{G-S}(u) - d_H(u) + e_H(u, S) \geq 1$, or

(2) $d_{G-S}(u) - d_H(u) + e_H(u, S) = 0$ and $d_H(u) - e_H(u, S) \geq 1$.

Since $\{x_1, x_2\} \cap V(H) \neq \emptyset$ and any vertex $v \in T \setminus (\{x_1, x_2\} \cup V(H))$ satisfies (1), we have

$$\sum_{x \in T}(d_{G-S}(x) - d_H(x) + e_H(x, S))$$
$$\geq |T| - 2 - 2m + 1 = |T| - 2m - 1. \tag{4.3.6}$$

Using (4.3.1), (4.3.2), (4.3.5), (4.3.6), $|S| + |T| \leq n$,
$$n \geq 9k - 1 - 4\sqrt{2(k-1)^2 + 2} + 2(2k+1)m$$

and Lemma 4.3.1, we obtain

$$-1 \geq k|S| + \sum_{x \in T}(d_{G-S}(x) - d_H(x) + e_H(x, S) - k)$$
$$\geq k|S| + |T| - 2m - 1 - k|T| = k|S| - (k-1)|T| - 2m - 1$$
$$\geq k|S| - (k-1)(n - |S|) - 2m - 1$$
$$= (2k-1)|S| - (k-1)n - 2m - 1$$
$$\geq (2k-1)\left(\frac{n+k-2}{2} - (d_H(x_1) + d_H(x_2) - e_H(x_1, S) - e_H(x_2, S))\right)$$
$$\quad - (k-1)n - 2m - 1$$
$$\geq (2k-1)\left(\frac{n+k-2}{2} - m\right) - (k-1)n - 2m - 1$$
$$= \frac{n}{2} + \frac{(2k-1)(k-2)}{2} - (2k+1)m - 1$$
$$\geq \frac{n}{2} - (2k+1)m - 1$$
$$\geq \frac{9k - 1 - 4\sqrt{2(k-1)^2 + 2} + 2(2k+1)m}{2} - (2k+1)m - 1$$

4.3 Neighborhood and Fractional (k, m)-Deleted Graphs

$$= \frac{9k - 1 - 4\sqrt{2(k-1)^2 + 2}}{2} - 1 > 0,$$

this is a contradiction.

Subcase 2.2 $1 \leq h_2 \leq k - 1$.

According to $d_H(x_1) \leq m$, we get

$$|N_T[x_1]| \leq d_{G-S}(x_1) + 1 = h_1 + d_H(x_1) - e_H(x_1, S) + 1$$
$$\leq h_1 + m + 1.$$

Complying this with (4.3.1), (4.3.2), (4.3.5), $m \geq 1$, $0 \leq h_1 \leq h_2 \leq k - 1$,

$$n \geq 9k - 1 - 4\sqrt{2(k-1)^2 + 2} + 2(2k+1)m$$

and $|S| + |T| \leq n$, we have

$$-1 \geq k|S| + \sum_{x \in T}(d_{G-S}(x) - d_H(x) + e_H(x, S) - k)$$

$$\geq k|S| + h_1|N_T[x_1]| + h_2(|T| - |N_T[x_1]|) - k|T|$$

$$= k|S| + (h_1 - h_2)|N_T[x_1]| + (h_2 - k)|T|$$

$$\geq k|S| + (h_1 - h_2)(h_1 + m + 1) + (h_2 - k)(n - |S|)$$

$$= (2k - h_2)|S| + (h_1 - h_2)(h_1 + m + 1) - (k - h_2)n$$

$$\geq (2k - h_2)\left(\frac{n + k - 2}{2} - h_1 - h_2 - m\right)$$
$$+ (h_1 - h_2)(h_1 + m + 1) - (k - h_2)n$$

$$= h_2^2 + \frac{n - 5k}{2}h_2 + h_1^2 + (m + 1 - 2k)h_1 + k(k - 2) - 2km$$

$$\geq h_2^2 + \frac{n - 5k}{2}h_2 + h_1^2 + (2 - 2k)h_1 + k(k - 2) - 2km$$

$$\geq h_2^2 + \frac{n - 5k}{2}h_2 + h_2^2 + (2 - 2k)h_2 + k(k - 2) - 2km$$

$$= 2h_2^2 + \frac{n - 9k + 4}{2}h_2 + k(k - 2) - 2km$$

$$\geq 2h_2^2 - 2\sqrt{2(k-1)^2 + 2}\,h_2 + (2k+1)mh_2 + \frac{3}{2}h_2$$
$$+ k(k - 2) - 2km$$

$$\geq 2h_2^2 - 2\sqrt{2(k-1)^2 + 2h_2} + (2k+1)m + \frac{3}{2}h_2$$
$$+ k(k-2) - 2km$$
$$\geq 2h_2^2 - 2\sqrt{2(k-1)^2 + 2h_2} + \frac{3}{2}h_2 + k(k-2) + 1$$
$$= \frac{1}{2}(2h_2 - \sqrt{2(k-1)^2 + 2})^2 + \frac{3}{2}h_2 - 1$$
$$\geq \frac{3}{2}h_2 - 1 \geq \frac{1}{2} > 0.$$

It is a contradiction. This completes the proof of Theorem 4.3.1. □

We obtain a new neighborhood condition for a graph to be a fractional (k,m)-deleted graph,[28] which is the following theorem.

Theorem 4.3.2 [28] *Let $k \geq 2$ and $m \geq 0$ be two integers, and let G be a graph of order n with*
$$n \geq 8k^2 + 4k - 8 + 8m(k+1) + \frac{4m-2}{k+m-1}.$$
If $\delta(G) \geq k + 2m$ and
$$|N_G(x) \cup N_G(y)| \geq \frac{n}{2}$$
for any two nonadjacent vertices x and y of G such that $N_G(x) \cap N_G(y) \neq \emptyset$, then G is a fractional (k,m)-deleted graph.

Proof Suppose that G satisfies the assumption of Theorem 4.3.2, but is not a fractional (k,m)-deleted graph. Then by Theorem 4.1.2, there exists some subset S of $V(G)$ such that
$$k|S| + \sum_{x \in T}(d_{G-S}(x) - d_H(x) + e_H(x,S) - k) \leq -1, \quad (4.3.7)$$
where $T = \{x : x \in V(G) \setminus S, d_{G-S}(x) - d_H(x) + e_H(x,S) \leq k-1\}$ and H is some subgraph of G with m edges. It is easy to see that
$$d_{G-S}(x) - d_H(x) + e_H(x,S) \geq 0$$
for each $x \in V(G)$. Since $|E(H)| = m$, we obtain

4.3 Neighborhood and Fractional (k, m)-Deleted Graphs

$$\sum_{x \in T} d_H(x) - e_H(S, T) \leq 2m.$$

Now, we prove the following claims.

Claim 1 $|T| \geq |S| + 1$.

Proof According to (4.3.7) and $d_{G-S}(x) - d_H(x) + e_H(x, S) \geq 0$ for each $x \in V(G)$, we get

$$-1 \geq k|S| + \sum_{x \in T}(d_{G-S}(x) - d_H(x) + e_H(x, S) - k) \geq k|S| - k|T|,$$

that is,

$$|T| \geq |S| + \frac{1}{k}.$$

In terms of the integrity of $|S|$ and $|T|$, we have $|T| \geq |S| + 1$. The proof of Claim 1 is complete. □

Claim 2 $|T| \geq k + m + 1$.

Proof If $|T| \leq k + m$, then by (4.3.7), Claim 1, $\delta(G) \geq k + 2m$ and $d_H(x) \leq m$ for each $x \in V(G)$, we get

$$-1 \geq k|S| + \sum_{x \in T}(d_{G-S}(x) - d_H(x) + e_H(x, S) - k)$$

$$= (k + m)|S| + \sum_{x \in T}(d_{G-S}(x) - d_H(x) + e_H(x, S) - k) - m|S|$$

$$\geq |T||S| + \sum_{x \in T}(d_{G-S}(x) - d_H(x) + e_H(x, S) - k) - m|S|$$

$$= \sum_{x \in T}(|S| + d_{G-S}(x) - d_H(x) + e_H(x, S) - k) - m|S|$$

$$\geq \sum_{x \in T}(\delta(G) - m - k) - m|S| \geq m|T| - m|S|$$

$$= m(|T| - |S|) \geq 0.$$

This is a contradiction. Hence, $|T| \geq k + m + 1$. This completes the proof of Claim 2. □

Claim 3 $1 \leq |S| < \frac{n}{2}$.

Proof If $S = \emptyset$, then by (4.3.7), $\delta(G) \geq k + 2m$ and $d_H(x) \leq m$ for each $x \in V(G)$, we obtain

$$-1 \geq \sum_{x \in T}(d_G(x) - d_H(x) - k) \geq \sum_{x \in T}(\delta(G) - m - k) \geq m|T| \geq 0,$$

a contradiction. Hence, $|S| \geq 1$.

On the other hand, from (4.3.7), $|S| + |T| \leq n$ and

$$d_{G-S}(x) - d_H(x) + e_H(x, S) \geq 0$$

for each $x \in V(G)$, we have

$$-1 \geq k|S| + \sum_{x \in T}(d_{G-S}(x) - d_H(x) + e_H(x, S) - k)$$

$$\geq k|S| - k|T| \geq k|S| - k(n - |S|)$$

$$= 2k|S| - kn,$$

which implies $|S| < \frac{n}{2}$. The proof of Claim 3 is complete. $\square$

Claim 4 $|S| < \frac{n}{2} - 2(k + m - 1)$.

Proof Suppose that $|S| \geq \frac{n}{2} - 2(k + m - 1)$. In terms of (4.3.7), $|S| + |T| \leq n$ and $\sum_{x \in T} d_H(x) - e_H(S, T) \leq 2m$, we obtain

$$-1 \geq k|S| + \sum_{x \in T}(d_{G-S}(x) - d_H(x) + e_H(x, S) - k)$$

$$\geq k|S| + \sum_{x \in T} d_{G-S}(x) - 2m - k|T|$$

$$\geq k|S| + \sum_{x \in T} d_{G-S}(x) - 2m - k(n - |S|)$$

$$= 2k|S| + \sum_{x \in T} d_{G-S}(x) - 2m - kn$$

$$\geq 2k\left(\frac{n}{2} - 2(k + m - 1)\right) + \sum_{x \in T} d_{G-S}(x) - 2m - kn$$

$$= \sum_{x \in T} d_{G-S}(x) - 4k(k + m - 1) - 2m,$$

4.3 Neighborhood and Fractional (k, m)-Deleted Graphs

which implies,

$$\sum_{x \in T} d_{G-S}(x) \leq 4k(k+m-1) + 2m - 1. \qquad (4.3.8)$$

According to (4.3.8), Claim 1 and

$$n \geq 8k^2 + 4k - 8 + 8m(k+1) + \frac{4m-2}{k+m-1},$$

we have

$$\frac{\sum_{x \in T} d_{G-S}(x)}{|T|} \leq \frac{4k(k+m-1) + 2m - 1}{|S| + 1}$$

$$\leq \frac{4k(k+m-1) + 2m - 1}{\frac{n}{2} - 2(k+m-1) + 1}$$

$$\leq 1 - \frac{1}{k+m}.$$

If follows from Claim 2 and the inequalities above that

$$\sum_{x \in T} d_{G-S}(x) \leq \left(1 - \frac{1}{k+m}\right)|T| = |T| - \frac{1}{k+m}|T|$$

$$< |T| - 1. \qquad (4.3.9)$$

Set $T_0 = \{x : x \in T, d_{G-S}(x) = 0\}$. Note that $|T_0| \geq 2$ holds by (4.3.9). For each two vertices $x, y \in T_0$, we get

$$|N_G(x) \cup N_G(y)| \leq |S| < \frac{n}{2}$$

by Claim 3. According to the definition of T_0, it is obvious that T_0 is an independent subset of G. Combining this with the assumption of Theorem 4.3.2, the neighborhoods of the vertices in T_0 are disjoint. Therefore, we get

$$|S| \geq \left|\bigcup_{x \in T_0} N_G(x)\right| \geq \delta(G)|T_0| \geq (k+2m)|T_0|. \qquad (4.3.10)$$

From (4.3.9) and the definition of T_0, we get

$$\left(1 - \frac{1}{k+m}\right)|T| \geq \sum_{x \in T} d_{G-S}(x) \geq |T| - |T_0|,$$

that is,
$$|T_0| \geq \frac{1}{k+m}|T|. \qquad (4.3.11)$$

Using (4.3.10) and (4.3.11), we obtain
$$|S| \geq (k+2m)|T_0| \geq \left(1 + \frac{m}{k+m}\right)|T| \geq |T|.$$

Which contradicts Claim 1. This completes the proof of Claim 4. □

Claim 5 $e_G(S,T) \leq (k+m)|S|$.

Proof According to Claim 2 and $d_{G-S}(x) \leq k+m-1$ for each $x \in T$, there exist at least two independent vertices $x, y \in T$. Moreover, by Claim 4 and $d_{G-S}(x) \leq k+m-1$ for each $x \in T$, we have
$$|N_G(x) \cup N_G(y)| \leq |S| + d_{G-S}(x) + d_{G-S}(y) < \frac{n}{2} \qquad (4.3.12)$$
for any two vertices $x, y \in T$.

In terms of (4.3.12) and the assumption of Theorem 4.3.2, $G[N_G(s) \cap T]$ is a complete induced subgraph of G for each $s \in S$. Note that $S \neq \emptyset$ by Claim 3. Thus, from $d_{G-S}(x) \leq k+m-1$ for each $x \in T$, we obtain
$$e_G(s,T) \leq \Delta(G[T]) + 1 \leq k+m.$$

Therefore, we get
$$e_G(S,T) \leq (k+m)|S|.$$

This completes the proof of Claim 5. □

By (4.3.7),
$$\sum_{x \in T} d_H(x) - e_H(S,T) \leq 2m, \quad \delta(G) \geq k+2m,$$

Claim 1, Claim 2 and Claim 5, we obtain
$$-1 \geq k|S| + \sum_{x \in T}(d_{G-S}(x) - d_H(x) + e_H(x,S) - k)$$
$$\geq k|S| + \sum_{x \in T}(d_{G-S}(x) - k) - 2m$$
$$= k|S| + \sum_{x \in T}(d_G(x) - k) - e_G(S,T) - 2m$$

4.3 Neighborhood and Fractional (k, m)-Deleted Graphs

$$\geq k|S| + \sum_{x \in T}(\delta(G) - k) - (k+m)|S| - 2m$$

$$\geq k|S| + (k + 2m - k)|T| - (k+m)|S| - 2m$$

$$= m(|T| - |S|) + m|T| - 2m$$

$$\geq 0,$$

it is a contradiction. Hence, Theorem 4.3.2 is proved. □

Remark 4.3.1 Let us show that the condition

$$|N_G(x) \cup N_G(y)| \geq \frac{n}{2}$$

in Theorem 4.3.2 can not be replaced by

$$|N_G(x) \cup N_G(y)| \geq \frac{n}{2} - 1.$$

We can show this by constructing a graph $G = ktK_1 \vee (kt+1)K_1$, where $k \geq 2$ and $m \geq 0$ are two integers and t is enough large positive integer. Then it follows that $|V(G)| = n = 2kt + 1$ and

$$\frac{n}{2} > |N_G(x) \cup N_G(y)| = kt > \frac{n}{2} - 1$$

for any two nonadjacent vertices x and y of $(kt+1)K_1$ such that $N_G(x) \cap N_G(y) \neq \emptyset$. Let

$$S = V(ktK_1) \subseteq V(G), \quad T = V((kt+1)K_1) \subseteq V(G)$$

and H is any subgraph of G with m edges. Then

$$|S| = kt, \quad |T| = kt + 1, \quad d_{G-S}(T) = 0$$

and $\sum_{x \in T} d_H(x) - e_H(S, T) = 0$. Thus, we obtain

$$k|S| + \sum_{x \in T}(d_{G-S}(x) - d_H(x) + e_H(x, S) - k)$$

$$= k^2 t - k(kt+1) = -k < 0.$$

According to Theorem 4.1.2, G is not a fractional (k, m)-deleted graph. In the sense above, the result of Theorem 4.3.2 is best possible.

By Theorem 4.3.2, it is clear that the following result holds.

Theorem 4.3.3 *Let $k \geq 2$ and $m \geq 0$ be two integers, and let G be a graph of order n with*

$$n \geq 8k^2 + 4k - 8 + 8m(k+1) + \frac{4m-2}{k+m-1}.$$

If $\delta(G) \geq k + 2m$ and

$$|N_G(x) \cup N_G(y)| \geq \frac{n}{2}$$

for each pair of nonadjacent vertices x, y of G, then G is a fractional (k, m)-deleted graph.

The result of Theorem 4.3.3 is stronger than one of Theorem 4.3.1 if $\delta(G) \geq k + 2m$ and the order n is sufficiently large. Set $m = 0$ in Theorem 4.3.2. Then we obtain the following result.

Theorem 4.3.4 *Let $k \geq 2$ be an integer, and let G be a graph of order n with*

$$n \geq 8k^2 + 4k - 8 - \frac{2}{k-1}.$$

If $\delta(G) \geq k$ and

$$|N_G(x) \cup N_G(y)| \geq \frac{n}{2}$$

for any two nonadjacent vertices x and y of G such that $N_G(x) \cap N_G(y) \neq \emptyset$, then G has a fractional k-factor.

4.4 Binding Number and Fractional (k, m)-Deleted Graphs

Binding number plays an important role in the research of factors and fractional factors in graphs. Many authors obtained the binding number condition for the existence of factors and fractional factors in graphs. In this section, we shall present a binding number condition for a graph to be a fractional (k, m)-deleted graph. The main result is the following theorem.

4.4 Binding Number and Fractional (k, m)-Deleted Graphs

Theorem 4.4.1 [30] *Let $k \geq 2$ and $m \geq 0$ be two integers, and let G be a graph of order n with*

$$n \geq 4k - 6 + \frac{2m}{k-1}.$$

If

$$\text{bind}(G) > \frac{(2k-1)(n-1)}{k(n-2) - 2m + 2},$$

then G is a fractional (k, m)-deleted graph.

Remark 4.4.1 Let us show that the condition

$$\text{bind}(G) > \frac{(2k-1)(n-1)}{k(n-2) - 2m + 2}$$

in Theorem 4.4.1 cannot be replaced by

$$\text{bind}(G) \geq \frac{(2k-1)(n-1)}{k(n-2) - 2m + 2}.$$

Let $k \geq 2$ and $0 \leq m \leq 2k-1$ be two integers such that m is odd and $\frac{(k+1)m}{k}$ is an integer, and let $l = \frac{2k+m-1}{2}$ and $m = 2k - 3 + \frac{(k+1)m}{k}$. We write

$$n = m + 2l = 4k - 4 + \frac{(2k+1)m}{k}.$$

Obviously, n is a positive integer. Let $G = K_m \vee lK_2$ and $X = V(lK_2)$. Then for any $x \in X$, $|N_G(X \setminus x)| = n - 1$. By the definition of $\text{bind}(G)$,

$$\text{bind}(G) = \frac{|N_G(X \setminus x)|}{|X \setminus x|} = \frac{n-1}{2l-1} = \frac{n-1}{2k+m-2}$$

$$= \frac{(2k-1)(n-1)}{k(n-2) - 2m + 2}.$$

Let

$$S = V(K_m) \subseteq V(G), \quad T = V(lK_2) \subseteq V(G)$$

and H is any subgraph of $G[T]$ with m edges. Then $|S| = m$, $|T| = 2l$ and

$$\sum_{x \in T} d_H(x) - e_H(S, T) = 2m.$$

Thus, we obtain

$$\delta_G(S,T) = k|S| - k|T| + d_{G-S}(T) = k|S| - k|T| + |T|$$
$$= k|S| - (k-1)|T| = km - 2(k-1)l$$
$$= k\left(2k - 3 + \frac{(k+1)m}{k}\right) - (k-1)(2k+m-1)$$
$$= 2m - 1 < 2m = \sum_{x \in T} d_H(x) - e_H(S,T).$$

In terms of Theorem 4.1.2, G is not a fractional (k,m)-deleted graph. In the above sense, the result in Theorem 4.4.1 is best possible.

In the following we prove Theorem 4.4.1, which depends on the following lemma.

Lemma 4.4.1 [21] *Let G be a graph of order n with $bind(G) > c$. Then*
$$\delta(G) > n - \frac{n-1}{c}.$$

Proof of Theorem 4.4.1 Suppose that G satisfies the conditions of Theorem 4.4.1, but is not a fractional (k,m)-deleted graph. According to Theorem 4.1.2 there exist disjoint subsets S and T of $V(G)$ such that

$$\delta_G(S,T) = k|S| + \sum_{x \in T} d_{G-S}(x) - k|T|$$
$$\leq \sum_{x \in T} d_H(x) - e_H(S,T) - 1, \tag{4.4.1}$$

where H is some subgraph of G with m edges. Since $|E(H)| = m$, we have
$$\sum_{x \in T} d_H(x) - e_H(S,T) \leq 2m.$$

Thus, according to (4.4.1) we obtain
$$\delta_G(S,T) = k|S| + \sum_{x \in T} d_{G-S}(x) - k|T| \leq 2m - 1. \tag{4.4.2}$$

We choose subsets S and T such that $|T|$ is minimum. Obviously, $T \neq \emptyset$ by (4.4.1).

Claim 1 $d_{G-S}(x) \leq k - 1$ for any $x \in T$.

4.4 Binding Number and Fractional (k, m)-Deleted Graphs

Proof If $d_{G-S}(x) \geq k$ for some $x \in T$, then the subsets S and $T\setminus\{x\}$ satisfy (4.4.2). This contradicts the choice of S and T. □

Set
$$h = \min\{d_{G-S}(x) : x \in T\},$$
and choose $x_1 \in T$ with $d_{G-S}(x_1) = h$. In terms of Claim 1, we have $0 \leq h \leq k-1$. Obviously, the following inequalities hold.
$$\delta(G) \leq d_G(x_1) \leq d_{G-S}(x_1) + |S| = h + |S|,$$
that is,
$$|S| \geq \delta(G) - h. \tag{4.4.3}$$

Using (4.4.3), Lemma 4.4.1 and $\operatorname{bind}(G) > \frac{(2k-1)(n-1)}{k(n-2)-2m+2}$, we have
$$|S| > \frac{(k-1)(n+2) + 2m}{2k-1} - h. \tag{4.4.4}$$

Now in order to prove the theorem, we shall deduce some contradictions by the following three cases.

Case 1 $2 \leq h \leq k-1$.

Subcase 1.1 $|T| < \dfrac{k(n-2) - 2m + 2}{2k-1} + h.$

In this case, it is easy to see that
$$|T| \leq \frac{k(n-2) - 2m + 1}{2k - 1} + h. \tag{4.4.5}$$

From (4.4.4), we obtain
$$|S| \geq \frac{(k-1)(n+2) + 2m + 1}{2k-1} - h. \tag{4.4.6}$$

In terms of (4.4.2), we have
$$-1 \geq \delta_G(S, T) - 2m = k|S| + \sum_{x \in T} d_{G-S}(x) - k|T| - 2m$$
$$\geq k|S| - (k-h)|T| - 2m. \tag{4.4.7}$$

Multiplying (4.4.7) by $2k - 1$ and rearranging, and then using (4.4.5) and (4.4.6),

$$0 \geq k(2k-1)|S| - (k-h)(2k-1)|T| - (2k-1)(2m-1)$$
$$\geq k((k-1)(n+2) + 2m + 1 - h(2k-1))$$
$$- (k-h)(k(n-2) - 2m + 1 + h(2k-1))$$
$$- (2k-1)(2m-1)$$
$$= (h-1)(kn - (2k-1)(2k-h) - 2m) + 2k - 1,$$

that is,

$$0 \geq (h-1)(kn - (2k-1)(2k-h) - 2m) + 2k - 1. \qquad (4.4.8)$$

Subcase 1.1.1 $h = 2$.

Using (4.4.8) and $n \geq 4k - 6 + \frac{2m}{k-1}$, we get

$$0 \geq (h-1)(kn - (2k-1)(2k-h) - 2m) + 2k - 1$$
$$= kn - (2k-1)(2k-2) - 2m + 2k - 1$$
$$\geq k\left(4k - 6 + \frac{2m}{k-1}\right) - (2k-1)(2k-2) - 2m + 2k - 1$$
$$\geq 2k - 3 \geq 1,$$

which is a contradiction.

Subcase 1.1.2 $3 \leq h \leq k - 1$.

In terms of (4.4.8) and $n \geq 4k - 6 + \frac{2m}{k-1}$, we have

$$0 \geq (h-1)(kn - (2k-1)(2k-h) - 2m) + 2k - 1$$
$$\geq (h-1)(kn - (2k-1)(2k-3) - 2m) + 2k - 1$$
$$\geq (h-1)\left(4k^2 - 6k + \frac{2km}{k-1} - (2k-1)(2k-3) - 2m\right)$$
$$+ 2k - 1$$
$$\geq (h-1)(2k-3) + 2k - 1 \geq 2(2k-3) + 2k - 1$$
$$= 6k - 7 > 0.$$

That is a contradiction.

Subcase 1.2 $|T| \geq \dfrac{k(n-2) - 2m + 2}{2k - 1} + h.$

4.4 Binding Number and Fractional (k, m)-Deleted Graphs

Set $Y = T - N_{G-S}(x_1)$. Note that $|N_{G-S}(x_1)| = d_{G-S}(x_1)$. Thus, we have

$$|Y| \geq |T| - d_{G-S}(x_1) = |T| - h \geq \frac{k(n-2) - 2m + 2}{2k - 1} > 0$$

and

$$N_G(Y) \neq V(G).$$

Combining these with the definition of $\text{bind}(G)$, we obtain

$$\text{bind}(G) \leq \frac{|N_G(Y)|}{|Y|} \leq \frac{n-1}{|T| - h} \leq \frac{(2k-1)(n-1)}{k(n-2) - 2m + 2}.$$

That contradicts

$$\text{bind}(G) > \frac{(2k-1)(n-1)}{k(n-2) - 2m + 2}.$$

Case 2 $h = 1$.

Using (4.4.4) and $|S| + |T| \leq n$, we have

$$\delta_G(S, T) = k|S| + \sum_{x \in T} d_{G-S}(x) - k|T|$$

$$\geq k|S| + |T| - k|T| = k|S| - (k-1)|T|$$

$$\geq k|S| - (k-1)(n - |S|)$$

$$= (2k-1)|S| - (k-1)n$$

$$> (k-1)(n+2) + 2m - (2k-1) - (k-1)n$$

$$= 2m - 1,$$

which contradicts (4.4.2).

Case 3 $h = 0$.

Put $\lambda = |\{x : x \in T, d_{G-S}(x) = 0\}|$ and $X = V(G) \setminus S$. Clearly, $\lambda \geq 1$ and $N_G(X) \neq V(G)$ since $h = 0$, and

$$|X| = |V(G) \setminus S| \geq |T| \geq 1.$$

Thus, by the definition of $\text{bind}(G)$ we have

$$\frac{|N_G(X)|}{|X|} \geq \text{bind}(G) > \frac{(2k-1)(n-1)}{k(n-2) - 2m + 2},$$

that is,
$$|N_G(X)| > \frac{(2k-1)(n-1)}{k(n-2)-2m+2}|X|. \qquad (4.4.9)$$

On the other hand, it is easy to see that
$$|N_G(X)| \le n - \lambda. \qquad (4.4.10)$$

From (4.4.9), (4.4.10) and $|X| = n - |S|$, we obtain
$$n - \lambda > \frac{(2k-1)(n-1)}{k(n-2)-2m+2}|X| = \frac{(2k-1)(n-1)}{k(n-2)-2m+2}(n - |S|),$$
which implies
$$|S| > n - \frac{(n-\lambda)(k(n-2)-2m+2)}{(2k-1)(n-1)}. \qquad (4.4.11)$$

According to (4.4.2) and $|S| + |T| \le n$, we have
$$2m - 1 \ge \delta_G(S,T) = k|S| + \sum_{x \in T} d_{G-S}(x) - k|T|$$
$$\ge k|S| + |T| - \lambda - k|T| = k|S| - (k-1)|T| - \lambda$$
$$\ge k|S| - (k-1)(n - |S|) - \lambda$$
$$= (2k-1)|S| - (k-1)n - \lambda,$$
which implies
$$|S| \le \frac{(k-1)n + 2m - 1 + \lambda}{2k-1} = n - \frac{kn - 2m + 1 - \lambda}{2k-1}. \qquad (4.4.12)$$

From (4.4.11) and (4.4.12), we obtain
$$(n - \lambda)(k(n-2) - 2m + 2) > (n-1)(kn - 2m + 1 - \lambda). \qquad (4.4.13)$$

If the LHS and RHS of (4.4.13) are denoted by A and B respectively, then (4.4.13) says that $A - B > 0$. But, after some rearranging, we find that
$$A - B = -(k-1)n - \lambda((k-1)n - 2k - 2m + 3)$$
$$- 2m + 1. \qquad (4.4.14)$$

Since $n \ge 4k - 6 + \frac{2m}{k-1}$ and $\lambda \ge 1$, it is clear that the expression in (4.4.14) is negative, and this contradicts (4.4.13).

From the contradictions we deduce that G is a fractional (k,m)-deleted graph. This completes the proof of Theorem 4.4.1. □

4.5 Toughness and Fractional (k, m)-Deleted Graphs

In the previous sections, we discussed the relationship between degree, neighborhood, binding number and the existence of fractional (k, m)-deleted graphs. In this section, we investigate the fractional (k, m)-deleted graphs in terms of toughness and obtain a toughness condition for fractional (k, m)-deleted graphs,[44] which is an extension of Liu and Zhang's previous result on fractional k-factors.[14]

Theorem 4.5.1 [44] *Let k and m be two nonnegative integers with $k \geq 2$. A graph G with $\delta(G) \geq k + 2m$ is a fractional (k, m)-deleted graph if its toughness*

$$t(G) \geq k + \frac{2m-1}{k}.$$

The following lemmas is very useful in the proof of Theorem 4.5.1.

Lemma 4.5.1 [5] *If a graph G is not complete, then*

$$t(G) \leq \frac{1}{2}\delta(G).$$

Lemma 4.5.2 [14] *Let G be a graph and let $H = G[T]$ such that $d_{G-S}(x) = k - 1$ for every $x \in V(H)$ and no component of H is isomorphic to K_k where $T \subseteq V(G)$ and $k \geq 2$. Then there exist an independent set I and the covering set $C = V(H) - I$ of H satisfying*

$$|V(H)| \leq \left(k - \frac{1}{k+1}\right)|I|$$

and

$$|C| \leq \left(k - 1 - \frac{1}{k+1}\right)|I|.$$

Lemma 4.5.3 [14] *Let G be a graph. Set $H = G[T]$ with $\delta(H) \geq 1$ and $1 \leq d_G(x) \leq k - 1$ for any $x \in V(H)$, where $T \subseteq V(G)$ and $k \geq 2$ is* an integer. Let $T_1, T_2, \cdots, T_{k-1}$ be a partition of $V(H)$ such that $d_G(x) = j$ for $\forall x \in T_j$ (where T_j may be empty sets), $j = 1, 2, \cdots, k-1$. Suppose that each component of H has at least one vertex of degree no more than $k-2$ in G. Then there exist a maximal independent set I and a covering set $C = V(H) - I$ of H such that

$$\sum_{j=1}^{k-1}(k-j)c_j \leq \sum_{j=1}^{k-1}(k-2)(k-j)i_j,$$

where $i_j = |I \cap T_j|$, $c_j = |C \cap T_j|$, $j = 1, 2, \cdots, k-1$.

Proof of Theorem 4.5.1 Suppose that G satisfies the assumption of Theorem 4.5.1, but is not a fractional (k, m)-deleted graph. According to Theorem 4.1.2 there exist disjoint subsets S and T of $V(G)$ such that

$$\delta_G(S, T) = k|S| + \sum_{x \in T} d_{G-S}(x) - k|T|$$

$$< \sum_{x \in T} d_R(x) - e_R(S, T), \qquad (4.5.1)$$

where R is some subgraph of G with m edges. Since $|E(R)| = m$, we have

$$\sum_{x \in T} d_R(x) - e_R(S, T) \leq 2m.$$

Thus, from (4.5.1) we obtain

$$k|T| - \sum_{x \in T} d_{G-S}(x) > k|S| - 2m. \qquad (4.5.2)$$

We choose subsets S and T such that $|T|$ is minimum subject to (4.5.1) and (4.5.2). Obviously, $T \neq \emptyset$ by (4.5.1).

Claim 1 $d_{G-S}(x) \leq k-1$ for any $x \in T$.

Proof If $d_{G-S}(x) \geq k$ for some $x \in T$, then the subsets S and $T \setminus \{x\}$ satisfy (4.5.2). This contradicts the choice of S and T. $\square$

Claim 2 $S \neq \emptyset$.

Proof If $S = \emptyset$, then by $\delta(G) \geq k + 2m$, we have

$$d_{G-S}(x) = d_G(x) \geq \delta(G) \geq k + 2m \geq k$$

4.5 Toughness and Fractional (k,m)-Deleted Graphs

for each $x \in T$. This contradicts Claim 1. □

Let λ be the number of the components of $H' = G[T]$ which are isomorphic to K_k and let $T_0 = \{x : x \in V(H'), d_{G-S}(x) = 0\}$. Let H be the subgraph obtained from $H' - T_0$ by deleting those λ components isomorphic to K_k.

We shall consider two cases according to the value of $|V(H)|$ and derive contradictions.

Case 1 $|V(H)| = 0$.

In view of (4.5.2), we obtain

$$k|T_0| + k\lambda > k|S| - 2m.$$

Combining this with Claim 2, we have

$$1 \leq |S| < |T_0| + \lambda + \frac{2m}{k}. \tag{4.5.3}$$

If $|T_0| + \lambda = 0$, then we have $|T_0| = \lambda = 0$. Thus, we obtain

$$|T| = |T_0| + \lambda|K_k| + |V(H)| = 0,$$

which contradicts $T \neq \emptyset$. In the following, we may assume that $|T_0| + \lambda \geq 1$. Note that $\omega(G - S) \geq |T_0| + \lambda$. Therefore, we obtain

$$\omega(G - S) \geq |T_0| + \lambda \geq 1.$$

Subcase 1.1 $\omega(G - S) = 1$.

Obviously, $|T_0| + \lambda = 1$. According to Claim 1, for any $x \in T$ we have

$$k - 1 + |S| \geq d_{G-S}(x) + |S| \geq d_G(x) \geq \delta(G). \tag{4.5.4}$$

Using (4.5.3), (4.5.4), Lemma 4.5.1 and $|T_0| + \lambda = 1$, we obtain

$$2t(G) \leq \delta(G) \leq k - 1 + |S| < k - 1 + |T_0| + \lambda + \frac{2m}{k} = k + \frac{2m}{k},$$

which implies

$$t(G) < \frac{1}{2}\left(k + \frac{2m}{k}\right) < k + \frac{2m - 1}{k},$$

which contradicts that

$$t(G) \geq k + \frac{2m - 1}{k}.$$

Subcase 1.2 $\omega(G-S) > 1$.

According to $\omega(G-S) \geq |T_0| + \lambda \geq 1$ and (4.5.3), we have

$$t(G) \leq \frac{|S|}{\omega(G-S)} < \frac{|T_0| + \lambda + \frac{2m}{k}}{|T_0| + \lambda} = 1 + \frac{2m}{k(|T_0| + \lambda)}$$

$$\leq 1 + \frac{2m}{k} < k + \frac{2m-1}{k},$$

which contradicts that

$$t(G) \geq k + \frac{2m-1}{k}.$$

Case 2 $|V(H)| > 0$.

Clearly, $\delta(H) \geq 1$. Set $H = H_1 \cup H_2$, where H_1 is the union of components of H which satisfy that $d_{G-S}(x) = k-1$ for each $x \in V(H_1)$ and $H_2 = H - H_1$. According to Lemma 4.5.2, there are a maximal independent set I_1 and the covering set $C_1 = V(H_1) - I_1$ in H_1 such that

$$|V(H_1)| \leq \left(k - \frac{1}{k+1}\right)|I_1| \tag{4.5.5}$$

and

$$|C_1| \leq \left(k - 1 - \frac{1}{k+1}\right)|I_1|. \tag{4.5.6}$$

It is obvious that $\delta(H_2) \geq 1$ and $\Delta(H_2) \leq k - 1$. Let

$$T_j = \{x : x \in V(H_2), d_{G-S}(x) = j\}$$

for $1 \leq j \leq k-1$. From the definitions of H and H_2, we can also see that each component of H_2 has at least one vertex of degree no more than $k-2$ in $G - S$. By Lemma 4.5.3, H_2 has a maximal independent set I_2 and the covering set $C_2 = V(H_2) - I_2$ such that

$$\sum_{j=1}^{k-1}(k-j)c_j \leq \sum_{j=1}^{k-1}(k-2)(k-j)i_j, \tag{4.5.7}$$

where $i_j = |I_2 \cap T_j|$, $c_j = |C_2 \cap T_j|$, $j = 1, 2, \cdots, k-1$. Let

$$W = G - (S \cup T), \quad U = S \cup C_1 \cup C_2 \cup (N_G(I_2) \cap V(W)).$$

Then since

4.5 Toughness and Fractional (k, m)-Deleted Graphs

$$|C_2| + |N_G(I_2) \cap V(W)| \leq \sum_{j=1}^{k-1} ji_j,$$

we obtain

$$|U| \leq |S| + |C_1| + \sum_{j=1}^{k-1} ji_j \qquad (4.5.8)$$

and

$$\omega(G - U) \geq |T_0| + \lambda + |I_1| + \sum_{j=1}^{k-1} i_j. \qquad (4.5.9)$$

Now we prove the following claim.

Claim 3 $|U| \geq t(G)\omega(G - U)$.

Proof Claim 3 obviously holds for $\omega(G - U) = 0$. If $\omega(G - U) = 1$, then by Lemma 4.5.1 for each $x \in T$ we have

$$|U| \geq d_{G-S}(x) + |S| \geq d_G(x) \geq \delta(G) \geq 2t(G) \geq t(G)\omega(G - U).$$

If $\omega(G - U) \geq 2$, then from the definition of $t(G)$, we obtain

$$|U| \geq t(G)\omega(G - U).$$

The proof of Claim 3 is complete. □

In view of (4.5.8), (4.5.9) and Claim 3, we obtain

$$|S| + |C_1| \geq t(G)\left(|T_0| + \lambda + |I_1| + \sum_{j=1}^{k-1} i_j\right) - \sum_{j=1}^{k-1} ji_j. \qquad (4.5.10)$$

According to (4.5.2), we have

$$k|T_0| + k\lambda + |V(H_1)| + \sum_{j=1}^{k-1}(k-j)i_j + \sum_{j=1}^{k-1}(k-j)c_j$$

$$> k|S| - 2m. \qquad (4.5.11)$$

Using (4.5.10) and (4.5.11), we obtain

$$k|T_0| + k\lambda + |V(H_1)| + \sum_{j=1}^{k-1}(k-j)i_j + \sum_{j=1}^{k-1}(k-j)c_j + k|C_1|$$

$$> kt(G)\left(|T_0| + \lambda + |I_1| + \sum_{j=1}^{k-1} i_j\right) - k\sum_{j=1}^{k-1} ji_j - 2m.$$

Combining this with $k(t(G)-1)(|T_0|+\lambda) \geq 0$, we have

$$\sum_{j=1}^{k-1}(k-j)c_j + |V(H_1)| + k|C_1|$$

$$> \sum_{j=1}^{k-1}(kt(G) - kj - k + j)i_j + kt(G)|I_1| - 2m. \qquad (4.5.12)$$

According to (4.5.5) and (4.5.6), we have

$$|V(H_1)| + k|C_1| \leq \left(k - \frac{1}{k+1}\right)|I_1| + k\left(k - 1 - \frac{1}{k+1}\right)|I_1|$$

$$= (k^2 - 1)|I_1|. \qquad (4.5.13)$$

In view of (4.5.7), (4.5.12) and (4.5.13), we obtain

$$\sum_{j=1}^{k-1}(k-2)(k-j)i_j + (k^2-1)|I_1|$$

$$> \sum_{j=1}^{k-1}(kt(G) - kj - k + j)i_j + kt(G)|I_1| - 2m. \qquad (4.5.14)$$

Note that

$$|I_2| = \sum_{j=1}^{k-1} i_j \geq 1. \qquad (4.5.15)$$

From (4.5.14) and (4.5.15), we obtain

$$\sum_{j=1}^{k-1}(k-2)(k-j)i_j + (k^2-1)|I_1|$$

$$> \sum_{j=1}^{k-1}(kt(G) - kj - k + j)i_j + kt(G)|I_1| - 2m\sum_{j=1}^{k-1} i_j$$

$$= \sum_{j=1}^{k-1}(kt(G) - kj - k + j - 2m)i_j + kt(G)|I_1|,$$

that is,

$$\sum_{j=1}^{k-1}(k-2)(k-j)i_j + (k^2-1)|I_1|$$

$$> \sum_{j=1}^{k-1}(kt(G) - kj - k + j - 2m)i_j + kt(G)|I_1|.$$

4.5 Toughness and Fractional (k,m)-Deleted Graphs

Thus at least one of the following two cases must hold.

Subcase 2.1 There exists at least one j such that
$$(k-2)(k-j) > kt(G) - kj - k + j - 2m,$$
which implies
$$kt(G) < k^2 - k + j + 2m.$$
Combining this with $j \leq k-1$, we have
$$t(G) < k + \frac{2m-k+j}{k} \leq k + \frac{2m-1}{k},$$
which contradicts that
$$t(G) \geq k + \frac{2m-1}{k}.$$

Subcase 2.2 $k^2 - 1 > kt(G)$.

According to $t(G) \geq k + \frac{2m-1}{k}$, we obtain
$$0 > kt(G) - (k^2 - 1) \geq k^2 + 2m - 1 - (k^2 - 1) = 2m \geq 0,$$
which is a contradiction.

From the argument above, we deduce the desired contradictions. Hence, G is a fractional (k,m)-deleted graph. The proof of Theorem 4.5.1 is complete. □

Remark 4.5.1 Let us show that the condition
$$t(G) \geq k + \frac{2m-1}{k}$$
in Theorem 4.5.1 is sharp. Let $k \geq 2$, $m \geq 0$ and $n \geq 1$ be integers. Consider a graph G constructed from $K_{n(k-1)}$, $(nk+1)K_{k-1}$ and $K_{(nk+1)(k-1)+2mn}$ as follows: let
$$V((nk+1)K_{k-1}) = \{x_1, x_2, \cdots, x_{(nk+1)(k-1)}\}$$
and $\{y_1, y_2, \cdots, y_{(nk+1)(k-1)}\} \subset V(K_{(nk+1)(k-1)+2mn})$. Set
$$V(G) = V(K_{n(k-1)}) \cup V((nk+1)K_{k-1}) \cup V(K_{(nk+1)(k-1)+2mn})$$
and
$$E(G) = E(K_{n(k-1)}) \cup E((nk+1)K_{k-1}) \cup E(K_{(nk+1)(k-1)+2mn})$$

$$\cup \{x_i y_i : i = 1, 2, \cdots, (nk+1)(k-1)\}$$
$$\cup \{u x_i : u \in V(K_{n(k-1)}), i = 1, 2, \cdots, (nk+1)(k-1)\}.$$

Let $U = (K_{(nk+1)(k-1)+2mn} - \{y_1\}) \cup \{x_1\} \cup K_{n(k-1)}$. Then
$$|U| = (nk+n+1)(k-1) + 2mn$$
and $\omega(G - U) = nk + 2$. This follows that
$$t(G) = \frac{(nk+n+1)(k-1) + 2mn}{nk+2} < k + \frac{2m-1}{k}$$
and $t(G) \to k + \frac{2m-1}{k}$ when $n \to \infty$. Let
$$S = V(K_{n(k-1)}), \quad T = V((nk+1)K_{k-1})$$
and H is any subgraph of $G[S]$ with m edges. Then $|S| = n(k-1)$, $|T| = (nk+1)(k-1)$, $d_{G-S}(T) = (nk+1)(k-1)(k-1)$ and
$$\sum_{x \in T} d_H(x) - e_H(S, T) = 0.$$

Thus, we have
$$\delta_G(S, T) = k|S| + d_{G-S}(T) - k|T|$$
$$= kn(k-1) + (nk+1)(k-1)(k-1) - k(nk+1)(k-1)$$
$$= -k + 1 < 0 = \sum_{x \in T} d_H(x) - e_H(S, T).$$

By Theorem 4.1.2, G is not a fractional (k, m)-deleted graph. In the sense above, the condition
$$t(G) \geq k + \frac{2m-1}{k}$$
in Theorem 4.5.1 is sharp.

If $m = 1$ in Theorem 4.5.1, then we obtain the following result.

Theorem 4.5.2 *Let k be a nonnegative integer with $k \geq 2$. A graph G with $\delta(G) \geq k + 2$ is a fractional k-deleted graph if its toughness*
$$t(G) \geq k + \frac{1}{k}.$$

Chapter 5
Fractional (g, f, m)-Deleted Graphs

In this chapter, we investigate fractional (g, f, m)-deleted graphs, which are natural generalizations of fractional k-deleted graphs, fractional (g, f)-deleted graphs and fractional (k, m)-deleted graphs. We show some sufficient conditions for graphs to be fractional (g, f, m)-deleted graphs.

5.1 Preliminary and Results

Let g and f be two integer-valued functions defined on $V(G)$ satisfying $0 \le g(x) \le f(x)$ for any $x \in V(G)$. Then a (g, f)-factor is a spanning subgraph F of G such that $g(x) \le d_F(x) \le f(x)$ for each $x \in V(G)$. Let $h : E(G) \to [0, 1]$ be a function. If

$$g(x) \le \sum_{e \ni x} h(e) \le f(x)$$

holds for any $x \in V(G)$, then we call $G[F_h]$ a **fractional (g, f)-factor** of G with indicator function h where $F_h = \{e \in E(G) : h(e) > 0\}$. A graph G is called a **fractional (g, f, m)-deleted graph** if G has a fractional (g, f)-factor excluding any m edges. A fractional (g, f, m)-deleted graph is a fractional (f, m)-deleted graph if $g(x) = f(x)$ for any $x \in V(G)$. A fractional (f, m)-deleted graph is a fractional (k, m)-deleted graph if $f(x) = k$ for any $x \in V(G)$. A fractional $(g, f, 1)$-deleted graph is simply called a **fractional (g, f)-deleted graph**.

We shall show the fractional (g, f)-factor theorem presented by Anstee[2]. Liu and Zhang[12] gave a simple proof of the theorem.

Theorem 5.1.1 [2, 12] *Let G be a graph. Then G has a fractional (g, f)-factor if and only if for every subset S of $V(G)$,*

$$\delta_G(S, T) = f(S) + d_{G-S}(T) - g(T) \geq 0,$$

where $T = \{x : x \in V(G) \setminus S, \ d_{G-S}(x) \leq g(x)\}$.

We use Theorem 5.1.1 to prove the following theorems in next section.

Theorem 5.1.2 [43] *Let G be a graph, and let g, f be two integer-valued functions defined on $V(G)$ satisfying $0 \leq g(x) \leq f(x)$ for any $x \in V(G)$. If*

$$g(x) \leq r(d_G(x) - m) \leq f(x) - rm$$

holds for any $x \in V(G)$, then G is a fractional (g, f, m)-deleted graph, where m is a nonnegative integer and $0 < r \leq 1$ is a real.

Theorem 5.1.3 [43] *Let G be a graph, and let g, f be two integer-valued functions defined on $V(G)$ such that $0 \leq g(x) \leq f(x)$ for any $x \in V(G)$. If*

$$f(x)(d_G(y) - m) \geq d_G(x)g(y)$$

holds for any $x, y \in V(G)$ with $x \neq y$, then G is a fractional (g, f, m)-deleted graph, where m is a nonnegative integer.

If $m = 0$ in Theorem 5.1.2 and Theorem 5.1.3, then we obtain the following corollaries.

Corollary 7 *Let G be a graph, and let g, f be two integer-valued functions defined on $V(G)$ satisfying $0 \leq g(x) \leq f(x)$ for any $x \in V(G)$. If*

$$g(x) \leq rd_G(x) \leq f(x)$$

holds for any $x \in V(G)$, then G contains a fractional (g, f)-factor, where $0 < r \leq 1$ is a real.

Corollary 8 [13] *Let G be a graph, and let g, f be two integer-valued functions defined on $V(G)$ such that $0 \leq g(x) \leq f(x)$ for any $x \in V(G)$. If*

$$f(x)d_G(y) \geq d_G(x)g(y)$$

holds for any $x, y \in V(G)$ with $x \neq y$, then G admits a fractional (g, f)-factor.

If $m = 0$ in Theorem 5.1.2 and Theorem 5.1.3, then we obtain the following corollaries.

Corollary 9 *Let G be a graph, and let g, f be two integer-valued functions defined on $V(G)$ satisfying $0 \leq g(x) \leq f(x)$ for any $x \in V(G)$. If*
$$g(x) \leq r(d_G(x) - 1) \leq f(x) - rm$$
holds for any $x \in V(G)$, then G is a fractional (g, f)-deleted graph, where $0 < r \leq 1$ is a real.

Corollary 10 [22] *Let G be a graph, and let g, f be two integer-valued functions defined on $V(G)$ such that $0 \leq g(x) \leq f(x)$ for any $x \in V(G)$. If*
$$f(x)(d_G(y) - 1) \geq d_G(x) g(y)$$
holds for any $x, y \in V(G)$ with $x \neq y$, then G is a fractional (g, f)-deleted graph.

5.2 Proof of Main Theorems

Proof of Theorem 5.1.2 Let $e_i = u_i v_i$, $i = 1, 2, \cdots, m$ be any m edges of G. Set $G' = G - e_1 - e_2 - \cdots - e_m$. Obviously, we have
$$d_G(x) \geq d_{G'}(x) \geq d_G(x) - m \tag{5.2.1}$$
for any $x \in V(G)$.

In order to prove Theorem 5.1.2, we only need to verify that G' has a fractional (g, f)-factor. In terms of Theorem 5.1.1, we only need to prove that
$$\delta_{G'}(S, T) = d_{G'-S}(T) - g(T) + f(S) \geq 0$$
holds for any $S \subseteq V(G')$, where
$$T = \{x : x \in V(G') \setminus S, d_{G'-S}(x) \leq g(x)\}.$$

According to $g(x) \leq r(d_G(x) - m)$ and (5.2.1), we obtain

$$g(x) \leq rd_{G'}(x). \tag{5.2.2}$$

By the conditions of the theorem, (5.2.1) and (5.2.2), we have

$$\begin{aligned}\delta_{G'}(S,T) &= d_{G'-S}(T) - g(T) + f(S) \\ &\geq d_{G'-S}(T) - rd_{G'}(T) + rd_G(S) \\ &\geq d_{G'-S}(T) - rd_{G'}(T) + rd_{G'}(S) \\ &= d_{G'-S}(T) + r(d_{G'}(S) - d_{G'}(T))\end{aligned}$$

for any $S \subseteq V(G')$.

Note that $d_{G'}(S) - d_{G'}(T) \geq -d_{G'-S}(T)$. Thus, we obtain

$$\delta_{G'}(S,T) \geq d_{G'-S}(T) - rd_{G'-S}(T) = (1-r)d_{G'-S}(T) \geq 0.$$

In terms of Theorem 5.1.1, G' has a fractional (g, f)-factor, i.e., G is a fractional (g, f, m)-deleted graph. This completes the proof of Theorem 5.1.2.

□

Proof of Theorem 5.1.3 Let $e_i = u_i v_i$, $i = 1, 2, \cdots, m$ be any m edges of G. Set $G' = G - e_1 - e_2 - \cdots - e_m$. It is easy to see that

$$d_G(x) \geq d_{G'}(x) \geq d_G(x) - m$$

holds for any $x \in V(G)$.

In order to prove Theorem 5.1.3, we only need to verify that G' has a fractional (g, f)-factor. According to Corollary 8, we only need to prove that

$$f(x)d_{G'}(y) \geq d_{G'}(x)g(y) \tag{5.2.3}$$

holds for any $x \in V(G)$ with $x \neq y$.

For any $x \in V(G)$ with $x \neq y$, we have

$$f(x)d_{G'}(y) \geq f(x)(d_G(y) - m) \geq d_G(x)g(y) \geq d_{G'}(x)g(y).$$

Hence, (5.2.3) holds. In terms of Corollary 8, G' has a fractional (g, f)-factor. And so, G is a fractional (g, f, m)-deleted graph. This completes the proof of Theorem 5.1.3.

□

References

[1] J. Akiyama, M. Kano, Factors and factorizations of graphs (Springer), 2011.

[2] R. P. Anstee, An algorithmic proof of Tutte's f-factor theorem, J. Alqorithms 6(1)(1985), 112-131.

[3] J. A. Bondy, U. S. R. Murty, Graph Theory with Applications, GTM-244, Berlin: Springer, 2008.

[4] Q. Bian, S. Zhou, Independence number, connectivity and fractional (g, f)-factors in graphs, Filomat, Accept.

[5] V. Chvátal, Tough graphs and Hamiltonian circuits, Discrete Mathematics 5(1973), 215-228.

[6] H. Enomoto, Toughness and the existence of k-factors III, Discrete Mathematics 189(1998), 277-282.

[7] T. Iida, T. Nishimura, Neighborhood conditions and k-factors, Tokyo Journal of Mathematics 20(2)(1997), 411-418.

[8] P. Katerinis and D. R. Woodall, Binding numbers of graphs and the existence of k-factors, Quart. J. Math. Oxford 38(2)(1987), 221-228.

[9] K. Kotani, Binding numbers of fractional k-deleted graphs, Proceedings of the Japan Academy, Ser. A, Mathematical Sciences 86(4)(2010), 85-88.

[10] Z. Li, G. Yan, X. Zhang, On fractional f-covered graphs, OR Trasactions 6(4)(2002), 65-68.

References

[11] Z. Li, G. Yan, X. Zhang, On fractional (g,f)-deleted graphs, Mathematica Applicata (China) 16(1)(2003), 148-154.

[12] G. Liu, L. Zhang, Fractional (g,f)-factors of graphs, Acta Mathematica Scientia 21B(4)(2001), 541-545.

[13] G. Liu, L. Zhang, Maximum fractional $(0,f)$-factors of graphs, Math. Appl. 13(1)(2000), 31-35.

[14] G. Liu, L. Zhang, Toughness and the existence of fractional k-factors of graphs, Discrete Mathematics 308(2008), 1741-1748.

[15] Y. Ma, G. Liu, Isolated toughness and existence of fractional factors in graphs, Acta Mathematicae Applicatae Sinica 26(2003), 133-140.

[16] T. Niessen, Nash-Williams conditions and the existence of k-factors, Ars Combinatoria 34(1992), 251-256.

[17] M. D. Plummer, Graph factors and factorization: 1985-2003: A survey, Discrete Mathematics 307(7-8)(2007), 791-821.

[18] E. R. Schirerman, D. H. Ullman, Fractional Graph Theory, New York, John Wiley and Sons, Inc., 1997.

[19] C. Wang, A degree condition for the existence of k-factors with prescribed properties, International Journal of Mathematics and Mathematical Sciences 6(2005), 863-873.

[20] D. R. Woodall, k-factors and neighborhoods of independent sets in graphs, J. London Math. Soc. 41(2)(1990), 385-392.

[21] D. R. Woodall, The binding number of a graph and its Anderson number, J. Combin. Theory ser. B 15(1973), 225-255.

[22] J. Yang, Y. Ma, G. Liu, fractional (g,f)-factors in graphs, Applied Mathematics A Journal of Chinese Universities 16(2001), 385-390.

[23] J. Yu, G. Liu, M. Ma, B. Cao, A degree condition for graphs to have fractional factors, Advances in Mathematics (China) 35(5)(2006), 621-628.

References

[24] Q. Yu, G. Liu, Graph factors and matching extensions (Springer), 2009.

[25] L. Zhang, G. Liu, Fractional k-factors of graphs, J. Sys. Sci. and Math. Scis. 21(1)(2001), 88-92.

[26] S. Zhou, A minimum degree condition of fractional (k,m)-deleted graphs, Comptes rendus Mathematique 347(21-22)(2009), 1223-1226.

[27] S. Zhou, A neighborhood condition for graphs to be fractional (k,m)-deleted graphs, Glasgow Mathematical Journal 52(1)(2010), 33-40.

[28] S. Zhou, A new neighborhood condition for graphs to be fractional (k,m)-deleted graphs, Applied Mathematics Letters 25(3)(2012), 509-513.

[29] S. Zhou, A result on fractional k-deleted graphs, Mathematica Scandinavica 106(1)(2010), 99-106.

[30] S. Zhou, A sufficient condition for graphs to be fractional (k,m)-deleted graphs, Applied Mathematics Letters 24(9)(2011), 1533-1538.

[31] S. Zhou, Binding number and minimum degree for the existence of fractional k-factors with prescribed properties, Utilitas Mathematica 87(2012), 123-129.

[32] S. Zhou, Binding numbers and fractional (g,f)-deleted graphs, Utilitas Mathematica 93(2014), 305-314.

[33] S. Zhou, Degree conditions and fractional k-factors of graphs, Bulletin of the Korean Mathematical Society 48(2)(2011), 353-363.

[34] S. Zhou, Degree conditions for graphs to be fractional k-covered graphs, Ars Combinatoria, volume 119, January 2015.

[35] S. Zhou, On fractional (k,m)-deleted graphs, Utilitas Mathematica 89(2012), 193-201.

[36] S. Zhou, Some new sufficient conditions for graphs to have fractional k-factors, International Journal of Computer Mathematics 88(3)(2011), 484-490.

[37] S. Zhou, Some results on fractional k-factors, Indian Journal of Pure and Applied Mathematics 40(2)(2009), 113-121.

[38] S. Zhou, Toughness and the existence of fractional k-factors, Mathematics in Practice and Theory 36(6)(2006), 255-260.

[39] S. Zhou, Q. Bian, An existence theorem on fractional deleted graphs, Periodica Mathematica Hungarica, Accept.

[40] S. Zhou, H. Liu, Neighborhood conditions and fractional k-factors, Bulletin of the Malaysian Mathematical Sciences Society 32(1)(2009), 37-45.

[41] S. Zhou, H. Liu, Y. Xu, A result on the existence of fractional k-factors in graphs, Utilitas Mathematica, volume 96, March 2015.

[42] S. Zhou, B. Pu, Y. Xu, Neighborhood and the existence of fractional k-factors of graphs, Bulletin of the Australian Mathematical Society 81(3)(2010), 473-480.

[43] S. Zhou, C. Shang, Some sufficient conditions with fractional (g, f)-factors in graphs, Chinese Journal of Engineering Mathematics 24(2)(2007), 329-333.

[44] S. Zhou, Z. Sun, H. Ye, A toughness condition for fractional (k, m)-deleted graphs, Information Processing Letters 113(8)(2013), 255-259.

[45] S. Zhou, Z. Xu, Z. Duan, Binding number and fractional k-factors of graphs, Ars Combinatoria 102(2011), 473-481.